Cambridge Elements ☰

Organizational Response to Climate Change
edited by
Aseem Prakash
University of Washington
Jennifer Hadden
University of Maryland
David Konisky
Indiana University
Matthew Potoski
UC Santa Barbara

EXPLAINING TRANSFORMATIVE CHANGE IN ASEAN AND EU CLIMATE POLICY

Multilevel Problems, Policies, and Politics

Charanpal Bal
Satya Wacana Christian University
David Coen
University College London
Julia Kreienkamp
University College London
Paramitaningrum
Bina Nusantara University
Tom Pegram
University College London

CAMBRIDGE
UNIVERSITY PRESS

CAMBRIDGE
UNIVERSITY PRESS

Shaftesbury Road, Cambridge CB2 8EA, United Kingdom

One Liberty Plaza, 20th Floor, New York, NY 10006, USA

477 Williamstown Road, Port Melbourne, VIC 3207, Australia

314–321, 3rd Floor, Plot 3, Splendor Forum, Jasola District Centre, New Delhi – 110025, India

103 Penang Road, #05–06/07, Visioncrest Commercial, Singapore 238467

Cambridge University Press is part of Cambridge University Press & Assessment, a department of the University of Cambridge.

We share the University's mission to contribute to society through the pursuit of education, learning and research at the highest international levels of excellence.

www.cambridge.org
Information on this title: www.cambridge.org/9781009395953

DOI: 10.1017/9781009395960

When citing this work, please include a reference to the DOI 10.1017/9781009395960

First published 2023

A catalogue record for this publication is available from the British Library.

ISBN 978-1-009-39595-3 Paperback
ISSN 2753-9342 (online)
ISSN 2753-9334 (print)

Cambridge University Press & Assessment has no responsibility for the persistence or accuracy of URLs for external or third-party internet websites referred to in this publication and does not guarantee that any content on such websites is, or will remain, accurate or appropriate.

Explaining Transformative Change in ASEAN and EU Climate Policy

Multilevel Problems, Policies, and Politics

Organizational Response to Climate Change

DOI: 10.1017/9781009395960
First published online: April 2023

Charanpal Bal
Satya Wacana Christian University

David Coen
University College London

Julia Kreienkamp
University College London

Paramitaningrum
Bina Nusantara University

Tom Pegram
University College London

Author for correspondence: David Coen, d.coen@ucl.ac.uk

Abstract: The Paris Agreement embodies a flexible approach to global cooperation, aimed at encouraging ever more ambitious climate action by a variety of players on all levels of governance. Regional organizations play an important role in mobilizing such action. This Element provides novel insights into the conditions under which policy entrepreneurs can bring about transformative policy change in regional settings, with a focus on the European Union (EU) and the Association of Southeast Asian Nations (ASEAN). It finds that opportunity structures in the EU have been conducive to successful climate-progressive policy entrepreneurship at several key junctures, but not consistently. In contrast, the ASEAN governance context provides few access points for non-elite interests, making it fiendishly difficult for policy entrepreneurs to push for substantive policy change in the face of powerful domestic veto players. This title is also available as Open Access on Cambridge Core.

Keywords: environmental governance, climate policy, multilevel, multiple streams, European Union, ASEAN, European Green Deal

ISBNs: 9781009395953 (PB), 9781009395960 (OC)
ISSNs: 2753-9342 (online), 2753-9334 (print)

Contents

1 Introduction

How do global and regional climate targets, rules, policies, and standards emerge and under which conditions are they effectively enabled within domestic political systems? When and how do national policy innovations diffuse and who are the principal actors involved? Climate governance under the United Nations Framework Convention on Climate Change (UNFCCC) is not a linear process of global-to-local policy transmission. Rather, it is a product of dynamic, multilevel interactions, with a broad range of diverse actors jostling to upload, download, resist, impose, shape, and evade or enforce compliance with rules, standards, and norms. This Element combines insights from the literatures on multilevel governance (MLG) and policy entrepreneurship to address the question: what explains the ability of climate policy entrepreneurs to achieve transformative policy change at the regional level, with a focus on the European Union (EU) and the Association of Southeast Asian Nations (ASEAN).

While much climate governance scholarship focuses on the dysfunctions of the intergovernmental level, this Element identifies regional organizations as an instructive domain of analysis because they sit neither at the "top" nor at the "bottom" of the global climate change regime, providing vital governance (regulatory) as well as metagovernance (steering) functions (Sørensen and Torfing 2009). We break new ground empirically by comparing governance arrangements in the European Union (EU), where supranational climate policy-making is most advanced, to those in the Association of Southeast Asian Nations (ASEAN), where regional cooperation on climate change remains limited. Although there are significant differences between the EU and ASEAN, both case studies point to the potential importance of linkages across global, regional, and national climate governance domains, with the nonbinding Paris Agreement premised upon setting boundary conditions for enabling decentralized action by a cast of actors, from regional organizations to firms, municipalities, and individuals (Harrison and Geyer 2019).

However, while cross-MLG linkages have, at several points in time, accelerated policymaking processes in the EU, they have struggled to advance transformative climate policy action within ASEAN. To shed light on the factors that have facilitated or impeded more ambitious, multilevel policymaking across these two regimes, the Element supplements an MLG lens with John Kingdon's (1984, 1995, 2003) influential multiple streams framework (MSF). While MLG accounts for the increasingly interdependent and nested nature of climate policymaking across levels of governance, innovative uses of MSF have provided powerful insight into the role of policy entrepreneurs and the structural conditions (problem perception, availability of policy tools, and political will) which will often determine their ability to secure transformative policy change (Herweg et al. 2017). While much of the climate scholarship has rightly focused on the UNFCCC regime and coordination among member states, we employ the MLG-MSF framework to illuminate the conditions under which climate policy goals are actually implemented within domestic jurisdictions and the scope for action by a wide variety of actors at all levels.

Through an original paired case comparison across the ASEAN and EU climate policy regimes, we focus on an underexplored but central intervening variable: cross-level interactions and their combined impact upon regional climate policy outcomes, which is the dependent variable for this study. More specifically, we are interested in exploring the conditions under which multi-level interactions facilitate higher aggregate ambition and step changes in policymaking. In so doing, we build upon cutting-edge research by Bernstein and Hoffmann (2019: 921), among others, who direct our attention to the complex multilevel and interdependent nature of carbon lock-in and the important observation that the threshold for change will often be determined by "local" characteristics of the "carbon lock-in trap." Our approach also resonates with Green (2020: 153) who argues that it is crucial that climate scholars pay more attention to climate obstructionists and the political con-flicts embedded in the task of diversifying away from carbon-intensive indus-trial growth models. Finally, our focus on non-incremental policy changes reflects concerns, shared by critical scholars and others, that the scale and urgency of the climate challenge requires interventions that are "transforma-tive and not merely ameliorative" (Eckersley 2020: 2). Our findings not only contribute to advancing insights in this field of scholarship, but will also be of interest to policymakers seeking to better understand and reform policy processes with a view to making existing governance arrangements more effective.

We find that the EU's uniquely advanced MLG structures provide multiple entry points for diverse policy entrepreneurs and windows of opportunity from "above" and "below." However, importantly, the diffusion of policymaking authority in the EU does not necessarily favor progressive climate agendas and interactions with EU member states do not necessarily translate into high ambition climate policies. To bring about transformative policy change, policy entrepreneurs must be able to couple the problem, policy, and politics streams. The European Commission – assisted by supportive governments and non-state actors – emerges as the key policy entrepreneur in this arena. A closer look at the MSF dynamics of the European Emissions Trading System (ETS), European climate policymaking after the global financial crisis, and the European Green Deal (EGD) serves to illustrate this claim.

Whereas supranational climate policymaking is well established in the EU, regional cooperation in the ASEAN region remains comparatively limited and interaction with the UNFCCC regime is only loosely coordinated. UNFCCC policy initiatives have often not accelerated climate action within ASEAN or its member states, with the ASEAN regional climate mechanism endowed with few institutional prerogatives and ASEAN member states (AMS) beset by politico-business blockages, low state technical capacity, and elite preferences to progress premised on a carbon-intensive growth model. We argue that variation in policy outcomes is rooted not only in historically different international obligations, economic development levels or a distinct "ASEAN Way" of regional integration, but must also be understood in terms of domestic policy processes. Unlike the EU, where policy equilibria are constantly shifting, policy innovation at the ASEAN level has been stymied by resistance from powerful domestic politico-business coalitions, leaving few access points for non-elite policy entrepreneurs. We substantiate this argument through the examples of the ASEAN Agreement on Transboundary Haze Pollution (AATHP) and the ASEAN Power Grid (APG).

Data for this comparative study was largely collected from secondary and gray sources including government, industry, media, think tank, and NGO reports. A total of twenty-three targeted interviews were conducted with key actors within the ASEAN and EU, as well as industry bodies, think tanks, and NGOs in order to fill in key data gaps. For the EU case study, this included one from the UNFCCC Secretariat, one from EU Commission, one from the European Environment and Sustainable Development Advisory Councils, six EU government representatives (United Kingdom, Denmark, Finland), and six stakeholders (environmental INGOs, academics). The authors interviewed eight key actors for the ASEAN case study: two from industry bodies, and six stakeholders (environmental NGOs, think tanks, and academics). While the original project design envisaged a greater number of interviews with key stakeholders, the COVID-19 pandemic severely curtailed fieldwork plans. In some cases, interviews have been anonymized to protect the identity of the respondent.

This Element begins by introducing MLG and the MSF, which provide the theoretical framework for our case studies. We then identify the mechanisms by which transformative policy change occurs and the policy entrepreneurs driving that change, locating regional policy processes within their MLG context. This is followed by our two case studies of climate policy outcomes in the EU and ASEAN. The study concludes by reflecting on the implications of this analysis for the future of global, regional, and national climate policy-making and governance more broadly.

2 Understanding Multilevel Governance Dynamics: Problems, Policies, and Politics

The reallocation of authority and functions upward, downward, and sideways to domains of governance outside the traditional policy space over recent decades is usefully captured by the concept of MLG (Hooghe and Marks 2003). For our purposes, MLG configurations present a far more dynamic, strategic governance environment than unitary government systems, creating novel opportunities for multilevel actor coalitions to shape policy in ways favorable to their own interests and agendas. However, we cannot assume that MLG environments will advance ambitious policy agendas. To shed light on the conditions under which transformative policy change is more or less likely within MLG systems we draw on an extensive scholarship using John Kingdon's (1984, 1995, 2003) MSF to inquire into when policy entrepreneurs matter and under what conditions they can effectuate ambitious climate policies.

2.1 Multilevel Governance of Climate Change

As a useful point of departure, global climate change governance can be defined generally as, "all purposeful mechanisms and measures aimed at steering social systems toward preventing, mitigating, or adapting to the risks posed by climate change, established and implemented by states or other authorities" (Jagers and Stripple 2003: 385). A focus on "all purposeful mechanisms and measures," as well as the implementation prerogatives of "states or other authorities," invites reflection on the actor interactions across levels of governance, and the MLG lens provides a clear view into how state and non-state actors are embedded within wider intergovernmental and/or transnational governance regimes and how these regimes are, in turn, shaped by their constituent actors.

While a first generation of MLG researchers focused primarily on the diffusion of authority upward, with the EU offering the most advanced example of states ceding power to supranational institutions (Marks 1993; Scharpf 1994), more recent applications of MLG look beyond European integration, seeking to understand the complex and dynamic relationships between governmental and nongovernmental actors within and across territorially bounded spaces (Rietig 2014; Jänicke 2017). This new generation of MLG scholarship is more attuned to the realities of climate change governance, challenging rigid distinctions between local, national, international, and transnational politics. Within this broader understanding of MLG, authority is not necessarily explicitly delegated through legalized intergovernmental forums, but rather is dispersed and is often the emergent consequence of catalyzing action through informal intergovernmental networks, as well as transnational public-private governance initiatives (Roger 2020; Westerwinter et al. 2021).

Although states remain "the key players in the MLG system" (Jänicke 2017: 113), and the best resourced and most legitimate actors in the formulation and enforcement of climate policies (Jordan and Huitema 2014), regional governance arrangements have emerged as important intermediation arenas for enabling global policy implementation in local settings, enjoying three potential advantages: smaller number of actors, opportunities for issue-linkages, and formal interfacing with both national and international governance systems (Betsill 2007). We build upon this MLG scholarship to assess whether MLG configurations can raise policy ambition and enhance delivery effectiveness across diverse regional settings. More specifically, we bring into focus the role of agency in advancing or impeding policy goals across levels of governance, with particular focus on questions of location, focality, authority, and resources.

As a theoretical point of departure, it is helpful to distinguish between functionalist and post-functionalist MLG scholarship. The latter has emphasized the potential for actor coalitions to engage in positive "multilevel reinforcement" of best practices, taking advantage of efficiency gains through coordination and functional differentiation (Schreurs and Tibhergien 2007; Jänicke and Wurzel 2018). Progressive policy leaders may therefore find that they can promote their policy preferences across policy venues, attracting broader coalitions, and exploiting opportunities to causally induce policy spillover above or below (Rietig 2020). However, other contributions have challenged the functionalist optimism of much MLG literature, with criticism focusing primarily on questions of democratic legitimacy, transparency, and accountability (Pierre and Peters 2004; Papadopoulos 2010).

In a rapidly warming world, while policy innovation is vital, there is no reason to believe that MLG automatically generates incentives and opportunities for progressive policy entrepreneurship and "multilevel reinforcement," especially when it comes to implementation of policy agendas marked by new ideological cleavages, including environmentalism (Hooghe and Marks 2018). Complex MLG arrangements may further obscure accountability within domestic political systems, allowing powerful interests to capture the policymaking process (Curry 2015). In turn, the costs and benefits of climate action are not distributed equally, and disenfranchised groups may contest decarbonization policies if they are not socially inclusive (McCauley and Heffron 2018). As such, MLG arrangements must also contend with "policy obstructors" at the domestic level who are motivated to undermine or contest policies that threaten their interests (Hameiri and Jones 2017). Notably, in the case of climate change, potential policy obstructors are often able to mobilize significant resources, as exemplified by the average annual spending of €28 million (2010–2018) on EU lobbying activities by just five large fossil fuel companies and their industry groups (Laville 2019).

The hard reality of policymaking is, as Cairney and Zahariadis (2016: 87) observe, that policy change is reliant upon a "window of opportunity" during which "people pay attention to a problem, a viable solutions exists, and policy-makers have the motive and opportunity to select it." Such opportunities are often vanishingly rare. If positive multilevel reinforcement cannot be assumed, then under what conditions are policy entrepreneurs more or less likely to enable transformative climate action within MLG systems?

2.2 Mobilizing for Policy Change: The Multiple Streams Framework

To answer the above question, this Element builds on an extensive scholarship drawing on John Kingdon's (1984, 1995, 2003) MSF to inquire into when policy entrepreneurs matter and under what conditions they can effectuate action on ambitious climate standards and bring about policy change. A revision of the "garbage can model" of organizational decision-making (Cohen et al. 1972), Kingdon identifies three independent but frequently overlapping "streams" that inform policymaking processes:

1) *Problem stream:* issues arise that are deemed to require policy action (such as inequality, crime or poverty) because new evidence, crises, or public mobilization draw the attention of policymakers to the issue and convince them that they "should do something about" it (Kingdon 1995: 109).

2) *Policy stream:* potential policy solutions to these issues are developed, with ideas floating around in a "policy primeval soup," where they evolve as various actors seek to imprint their preferences guided by questions of technical feasibility, anticipation of future constraints, and normative acceptability (Kingdon 1995: 140–141).

3) *Politics stream:* changes in national mood, election outcomes, administrative turnover, or pressure group campaigns may all influence how receptive decision-makers are to proposed solutions, taking into account changing societal demands over time.

Policy change occurs if and when these three streams converge, thus creating a "window of opportunity." It is at this moment, that policy entrepreneurs have an opportunity to push forward their respective ideas, "coupling solutions to problems" and "both problems and solutions to politics" (Kingdon 1984: 21). Policy entrepreneurs assume a central causal function in the MSF as "advocates who are willing to invest their resources – time, energy, reputation, money – to promote a position in return for anticipated future gains in the form of material, purposive or solidary benefits" (Kingdon 1995: 179). More precisely, policy entrepreneurs work to couple these three relatively independent "streams" to achieve their desired ends. Alternatively, as "power brokers and manipulators" policy entrepreneurs may equally apply themselves to preventing such coupling from taking place (Zahariadis 2007: 74). Importantly, individual policy entrepreneurs are rarely able to achieve this on their own and they frequently engage in collaborative efforts and resource pooling to promote policy innovations in and around government (Mintrom 2019).

By highlighting opportunity structures, as well as the importance of agency, the MSF usefully emphasizes the interactive, strategic, and contingent nature of policymaking and its effect across venues. It also flags the importance of seizing the moment, given that "[t]he window in the first area opens up windows in adjacent areas, but they close rapidly as well" (Kingdon 2003: 192). In our analysis, we suggest that regional intermediation mechanisms can alter the duration of the policy reform window, impacting policymaking at other levels. However, this will depend upon the ability of those agencies which populate the regional mechanism being in a position to assume the role of an – at least partially – autonomous policy entrepreneur. While this is clearly the case for EU institutions such as the Commission, which enjoys special initiative rights, the remit of ASEAN's supranational institutional structures is carefully constrained.

"Coupling" is the central mechanism in the MSF, connecting the three streams to achieve policy change, stasis, or reversal. In an influential study, Zahariadis (2008: 520–525) advances four conditions to explain successful coupling within the EU regime: (1) entrepreneurial effectiveness, (2) framing the policy proposal to fit the preferred solution of policymakers, (3) strategic venue-shopping on agenda-setting and decision-coupling, and (4) the policymaking mode which may induce more or less agreement and conflict. Refining this argument further, Rietig (2020: 59) argues that EU-level policy coupling is more likely where the conditions "multi-level reinforcing dynamics" are present, including:

1. Interdependence between governance levels,
2. recognition that problems attached to one or more levels require policy solutions from a different level (or predetermined policies require problematizing from another level to gain political momentum), and
3. sufficient ambiguity to allow for venue shopping by policy entrepreneurs to seize opportunities of joining streams across levels and making use of open policy windows regardless of on which level they currently occur.

For our purposes, both Zahariadis and Rietig provide useful coordinates for our study with the former introducing the important distinction between "agenda-setting" and "decision-coupling" as consequential for successful coupling. However, it is important to flag that while such policy windows may be necessary, they are not sufficient to produce successful coupling. However skilled and dedicated the policy entrepreneur, policy outcomes will ultimately be determined by the politics stream (Kingdon 1995: 173). As Palmer (2015: 284) observes "events and conditions in the politics stream ... [are] essential in enabling observed policy outcomes." As such, how these actors interact with other variables in the policy process is likely to be key to placing their preferred outcome high on the decision-makers agenda (Palmer 2015: 272). In particular, success will hinge not only on the location of the entrepreneur across streams, but also the situational context and institutional environment that they encounter (Mukherjee and Howlett 2015). As the next section details, such considerations take on additional salience in MLG settings.

2.3 Understanding Policy Processes in Multilevel Governance

Employing an MSF approach helps advance the study of MLG in several ways. First, it shifts attention beyond intergovernmental multilateralism, broadening the scope of processes to be studied. Second, it makes visible the policy linkages across levels of governance, which supplement and overlay formal institutional structures. Third, it also brings to the fore questions of agency, providing a framework through which to assess when, how and why opportunities for successful policy entrepreneurship arise. Despite the analytical challenge posed by multilevel polities, scholars continue to refine the framework's application to globalized, ambiguous, and contested policy environments, leading Zohlnhöfer et al. (2015: 412) to conclude that "the MSF seems to have become more relevant and suitable than ever before."

Methodologically, for our purposes, the MSF focus on the interplay of the three independent streams provides a valuable organizing device to help frame a historical policy narrative (Zahariadis 2007: 81–82). Becker (2019: 149) concurs, arguing that the MSF provides "a comprehensible structure of these simultaneous processes," making possible tractable analysis of the interaction between agenda-setting and decision-making within MLG settings. However, as Ackrill and Kay (2011: 2) emphasize, it also identifies the key explanatory factor underlying policy outcomes, namely the "temporal conjunction of separate sub-policy processes." Simply put, when the three streams converge, this greatly increases the probability of policy agreement. Conversely, active efforts to prevent such convergence will likely have the opposite effect.

Of course, transferring the MSF approach to an MLG reality requires modification to take into account the implication for policy outcomes of processes (problem, policy, and politics streams) playing out at different territorial levels, subject to multi-stream brokering and advocacy by a far more diverse set of policy entrepreneurs. In Kingdon's original MSF, policy entrepreneurs are individuals. Importantly, however, these individuals rarely act on their own and networked collaborative efforts are fundamental to goal achievement (Mintrom 2019). Indeed, recent applications of the MSF have demonstrated how collective entities also act as policy entrepreneurs, from corporations (He and Ma 2019), nongovernmental organizations (Carter and Childs 2018), to supranational bodies such as the European Commission (Copeland and James 2013). In MLG settings, national governments can also be conceptualized as policy entrepreneurs, pursuing policy change on the regional or global level (Harcourt 2016).

We build upon existing scholarship to bring into sharper focus not only the multilevel reinforcing dynamics between the regional and the international, but also between the regional and the local. We also qualify the largely rational functionalist orientation of much of the MSF scholarship, emphasizing other factors which explain when, how, and why opportunities for successful policy entrepreneurship arise. While much of the MSF scholarship follows Ackrill and Kay (2011) in emphasizing informational advantage, for example, with a particular focus on policy framing "so as to provide [others with] guideposts for knowing, analyzing, persuading and acting" (Palmer 2015: 273), our MSF account of policy outcomes across MLG settings places significant weight on political economy factors, especially the role of powerful coalitions who often exercise decisive influence over the politics stream at the local level, and therefore over the crucial site of decision-coupling. Our comparative case studies suggest that strong links between domestic political decision-makers and business elites in high-emission sectors have been a key factor in preventing transformative policy change in ASEAN. Similar obstacles also arise in the EU, although political economy structures in Europe are, in comparison, more open to challenges by non-elite actors. As such, we are less sanguine on the prospects for securing "more ambitious policy outcomes despite temporary setbacks on some governance levels" (Rietig 2020: 56).

3 Comparing the EU and ASEAN: Levels, Streams, and Climate Policy

Whereas in Europe, regionalism is often associated with the rise of a single regional organization, the EU, Asia presents more of a patchwork of regional organizations that exist within regional boundaries. Nevertheless, for some, ASEAN provides an incipient move toward a more integrated regional regime complex (Yukawa 2018). As scholars have observed, the EU presents a "case of deep, supranational sovereignty-pooling" whereas ASEAN "is an example of distinctly sovereignty-friendly intergovernmental cooperation" (Larik 2019: 447). Normatively grounded in the principle of noninterference, ASEAN integration has not produced strong, independent institutions comparable to EU-level bodies such as the European Commission or the European Parliament.

The EU has long been a test bed for MLG application, with the concept serving to capture a uniquely complex multilevel political system which is governed by a multiplicity of actors and processes, beyond core intergovernmental venues. Compared to other regional organizations, the EU is uniquely autonomous as member states have delegated increasing executive, judicial, and legislative powers to European institutions (Pollack 2003). The high degree of regional integration has also allowed the EU to enhance its external "actorness" vis-à-vis multilateral venues such as the UNFCCC. However, while the EU is often lauded as highly pluralistic when it comes to EU policy networks, recent events lend credence to the claim that EU policy processes offer "more accountability, but less democracy" with little sign of broadscale societal participation within EU decision-making forums (Papadopoulos 2010: 1044). Indeed, as EU competencies have gradually expanded, so has political contestation, replacing the "permissive consensus" that initially allowed European integration to advance largely as a technocratic project with a "constraining dissensus" (Hooghe and Marks 2008).

Such considerations become acute when surveying other regional political systems. Yet, much of the MLG and MSF research agenda has principally focused on the United States or the EU, so it is important to carefully consider how to adapt these theories to our non-Western case study of ASEAN. Moreover, moving away from industrialized countries to more adverse political economy contexts, one could hypothesize that strong ties on the politics stream between policy entrepreneurs and the domestic executive might be paramount for effective policy change.

Regional governance in ASEAN is based on setting out rules and regulations that governments then have discretion to enact in domestic governance. However, ASEAN does not "pool sovereignty" to determine policy at the regional or domestic levels. Unlike in Europe, MLG has never been embraced by ASEAN governments as "truly modern governance" or as a "normative standard by which ... governance is presented as legitimate" (Jeffrey and Peterson 2020: 756). Rather, regional governance frameworks function minimally to (1) selectively "download" metagovernance norms from various global platforms; (2) set out regulatory guidelines and tasks for national governments to pursue their own objectives; and (3) provide support for knowledge-sharing and capacity-building in these areas. Frameworks such as the Asia Pacific Trade Agreement or the Council for Security Cooperation in the Asia Pacific have facilitated a reasonable level of regional coordination for policy transfer and diffusion. However, the efficacy of such initiatives depends in large part on the willingness of national and sub-national authorities to adopt and implement them.

As we demonstrate in our case studies, differences in both MLG structures and MSF dynamics account for variations in climate policy outcomes across the EU and ASEAN, notably the lack of non-incremental policy changes in the ASEAN case. An MLG-adjusted model of the MSF can also help us make sense of temporal fluctuations of ambition *within* regional settings, including periods of relative policy stagnation in the EU. We argue that variation in transformative policy outcomes, both within and across regional governance arrangements, is largely predicated on the degree of formal and informal coupling between agenda-setting and decision-making arenas. We now elaborate on this conceptual argument in more detail.

3.1 Multi-Stream Governance in Context: Agenda-Setting to Decision-Coupling

A basic task of adapting the MSF to EU and ASEAN policymaking is to identify and differentiate various actors, their location across streams, and their specific function during the critical "window" phases of policymaking on different levels of governance. More precisely, it also means separating out agenda-setting and decision-making as two distinct coupling phases with different windows of opportunity and constellations of actors and institutions. Our modified MSF model is illustrated in Figure 1 and serves as the coordinates for our comparative study. While interdependencies across governance levels can, under the right conditions, unleash reinforcing dynamics conducive to policy change, we argue that policy change – especially transformative policy change – will be dependent upon timing, context and, ultimately, agency within the political stream. We now clarify some of the key drivers and mechanisms of policy change which inform our case studies.

The MSF operates under constraints of ambiguity which are amplified in MLG settings. As Rietig (2020: 59) observes, streams running in parallel at different levels of governance can increase ambiguity to the advantage of policy entrepreneurs if they are able "to attach national-level solutions to international-level problems ... and vice versa." Conversely, entrepreneurs dedicated to blocking policy change may exploit ambiguity to decouple problems and solutions at different levels. In this respect, it is useful to distinguish between two forms of ambiguity in the MSF scholarship, ambiguity "as problematic preferences and

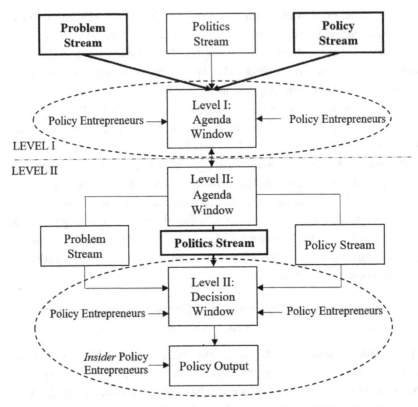

Figure 1 A modified multiple streams frameworks (MSF), adapted to a multilevel governance context (MLG)

unclear information about means ends" (Zohlnhöfer et al. 2015: 412) and "institutional ambiguity," referring to "a policy-making environment of overlapping institutions lacking a clear hierarchy" (Ackrill and Kay 2011: 5). As such, following Cairney and Jones (2016: 45), we are not just interested in how ambiguity relates to problem definition but also to policymaking responsibility at different levels of governance – presenting different levers to entrepreneurs intent on prizing open or slamming shut policy windows.

It is necessary to separate out both agenda-coupling and decision-coupling. Many MSF scholars have built upon Zahariadis' (1992, 2003) observation that agenda-setting is not the same as decision-making. We adapt the MSF to address linkages across levels and phases in the policymaking process, from the international to regional and sub-national. As Herweg et al. (2015: 444) observe, agenda-setting which results in "a worked out proposal ready for decision" ("Agenda Window") is different from "bargaining about the concrete design of the policy proposal" ("Decision Window"). Separation of policymaking phases also raises the question of unit of analysis, with this study applying the MSF not only to regional-level policy, but also adapting the framework to the sub-national level. We highlight, in particular, the role of regional-level organizations in shaping the policy and problem stream toward convergence on the Agenda Windows, while insider policy entrepreneurs often play an outsized role in the political stream which connects the Agenda Windows to the Decision Window (see Figure 1).

The political stream is dominant but not determinative. Kingdon (2011: 152) himself argued that "the application of the resources of major interest groups against a proposal does not necessarily carry the day," which, as Herweg et al. (2015: 438) rightly point out raises the question under what conditions will influential interest groups not carry the day? Is policy change or stasis certain if such groups are united in their preferences? Much ultimately depends upon the "openness" and "checks and balances" displayed by the political and institutional contexts in which policy-making is conducted. It is likely that policy entrepreneurs aligned with key veto players within domestic Level II political systems are likely to enjoy a structural advantage over Level I actors (Hameiri and Jones 2015). A dominant political stream is particularly likely in political systems where policy formation is ideo-logical and driven by "a [preconceived] political decision to pursue a particular policy 'solution'" (Howlett et al. 2015: 428). This is not to say that the problem and policy stream cannot catalyze windows to be opened. However, the political stream is likely to hold the key in terms of translating opportunities for agenda change (Agenda Windows) into actual policy change (Decision Window). Simply put, if decision-coupling is successful, then new policy is enacted.

Not all policy entrepreneurs are made alike. While supranational actors such as the Commission are often held up as entrepreneurs, due to its ability to initiate legislation and policy changes (Laffan 1997), there is no guarantee that entrepre-neurs at Level I will necessarily generate consensus at Level II, which is largely determined by power relations within domestic political systems. As such, we argue that while the supranational level may exercise considerable influence over analysis and framing of the collective problem, as well as link problem conver-gence to specific solutions, it exercises less compelling authority over the politics stream, especially when the focus moves to the Decision Window at the domestic level. As Herweg et al. (2015: 446) point out "insider political entrepreneurs" who can "draw on authority qua position" are best placed to couple streams. Ultimately, member states must sell those political bargains back home. And as Becker (2019: 151) notes, interested parties within domestic systems may inter-pret the criteria, technical feasibility, and normative acceptance of the specific solution quite differently to the regional-level policy stream.

Policy entrepreneurs may be necessary but they are not sufficient to cause policy change. They may succeed in coupling all three streams, and still fall at the final hurdle. As Ackrill et al. (2013: 879) observe, such fortuitous use of the policy window to couple streams "will never be enough to cause policy reform: we always require an account of the context." Achieving policy change is challenging in any context; the significance of variation in the political, institutional, and economic task environment takes on particular salience in this study which surveys MSF processes across two very different political systems. The ASEAN experience in this study provides a rare application of the MSF to policy processes outside Western developed countries and, as we explore, at the close, points to some important limitations with the original MSF framework which assumes that the politics stream is more dynamic and subject to competing pressures than is often the case.

4 Situating EU and ASEAN in Global Climate Governance

As we explore in this Element, policy entrepreneurs in the EU and ASEAN encounter very different opportunity structures, both in terms of MLG linkages (e.g., regional-level interaction with the UNFCCC) and in terms of the openness, accessibility, and fluidity of the three MSF streams. This, we argue, has important implications for the ability of policy entrepreneurs to push for policy change. More specifically, our analysis focuses on the kind of policy outcomes that imply a significant, non-incremental departure from previous policies with regard to direction, scale, and/or speed. These outcomes can be described as transformative if they have the potential to catalyze wider social, economic, and political change in line with ambitious climate objectives.

Bringing about such transformative change is difficult in any context. As Cairney (2018: 202) reminds us, windows of opportunities for major policy change open rarely and can "best be described as akin to a space launch in which policymakers will abort the mission unless every relevant factor is just right." Nevertheless, in the EU, MLG interactions have, at several points in time, facilitated the adoption of transformative policy innovations on climate change. This has included, for example, the inception of the first regional ETS or the adoption of the first Climate and Energy Package, both of which provided critical junctures in terms of defining preferred policy instruments and firmly establishing climate change as a major cross-cutting issue on the EU policy agenda. More recently, the EGD has the potential of being a pathway-defining moment, aiming to put the EU on a long-term trajectory to climate neutrality by 2050. ASEAN has also seen notable policy innovations, such as the APG initiative and the Agreement on Transboundary Haze Pollution; however, neither of these instruments has been primarily designed to serve climate mitigation targets and their transformational potential has been severely stymied by the selective engagement of national governments.

In short, while the EU has seen significant, and occasionally transformative, policy change as a result of multilevel interactions under the UNFCCC as well as the activities of various policy entrepreneurs, ASEAN is yet to see such path-defining moments. While long-standing differences in terms of normative preferences and institutional setup ("sovereignty pooling" vs. noninterventionism) play an important role in this regard, they do not fully explain ASEAN's lack of ambitious regional climate policies. Rather, to gain a more nuanced understanding of the conditions that enhance or restrict the potential for transformative policy change, we need to pay closer attention to the opportunity structures climate-progressive policy entrepreneurs encounter in regional settings, especially when it comes to connecting Agenda and Decision Windows across the three streams and across supranational (Level I) and domestic political systems (Level II).

This chapter serves to provide important context for our detailed EU and ASEAN cases studies. Below we provide key comparative data, for example, on population, gross domestic product (GDP), and greenhouse gas (GHG) emissions, followed by a brief discussion of how the EU and ASEAN have positioned themselves in the global climate governance regime and how they have sought to facilitate regional-level collaboration on climate change. We also offer a summary overview of our case studies, detailing how opportunity structures for policy entrepreneurs differ across the three MSF streams.

As Figures 2 and 3 show, historically, the EU has been a major contributor to global warming; however, since 1990, its emissions have followed a downward trend. In contrast, the emissions of ASEAN states have increased dramatically over the past three decades. Indonesia, for instance, now emits more than any EU state – much of it stemming from deforestation and land degradation – with its emissions amounting to about 4 percent of the global total in 2018. At the same time, the per capita emissions of most ASEAN states, including Indonesia, remain significantly below EU average. A closer look at the country-differentiated data (available in the Appendix) also reveals significant variation *within* the EU and ASEAN across indicators such as per capita emissions, GDP per capita, and emissions trajectories since 1990.

In terms of multilevel policy engagement, the EU has long been recognized as a powerful player in global climate governance. From the very start, it has taken a joint, if not always coherent, approach toward UNFCCC negotiations. Its long-standing ambition to be an international climate leader has been closely linked to internal objectives, above all building support for the wider European MLG project. Europe was an early adopter of a formal multilevel climate

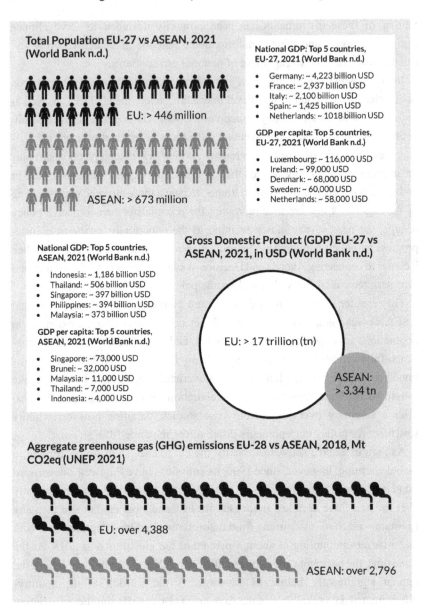

Figure 2 Population, GDP, and aggregate emissions – EU and ASEAN

architecture and has managed to establish a raft of internal institutional mechanisms as well as comparatively intrusive laws and regulations aimed at addressing climate change. The European Commission has played a key entrepreneurial role in this regard, leveraging its exclusive right to initiate legislation. However, the increasing cost of additional climate action policies has seen consensus among the twenty-seven member states fray at various points,

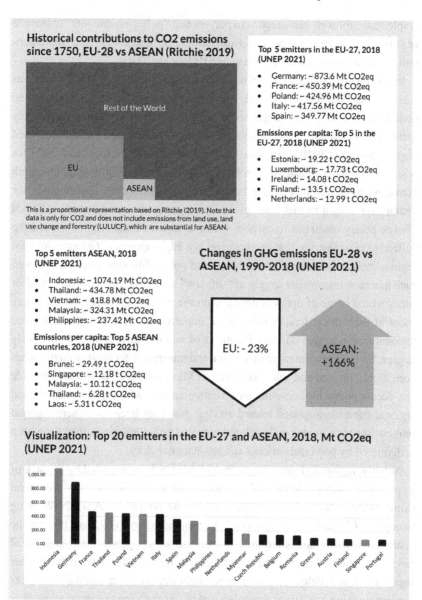

Figure 3 Historical emissions, relative changes in emissions, and top emitters – EU and ASEAN

sometimes pitting different levels and streams against one another. Climate policy is therefore characterized by a specific form of European MLG, which, while institutionalized, is also complex and often opaque as policymaking is conducted across territorial levels comprising large numbers of public and private actors (Marks 1996). As such, MLG arrangements in the EU provide ample space for well-resourced veto players and "policy obstructors" to impede ambitious environmental policies from passing or obstructing their implementation on the national level (Laffan and O'Mahony 2008).

In contrast to the EU, ASEAN has not engaged with the UNFCCC regime as a bloc. Under the Kyoto Protocol, AMS had no mitigation responsibilities and until quite recently, their engagement with the UNFCCC regime has largely focused on defending a narrow interpretation of the common but differentiated responsibilities (CBDR) principle. On the regional level, it took until 2007 for ASEAN to issue a Declaration on Environmental Sustainability at its thirteenth annual summit. As such, there have historically been few opportunities for institutionalizing a formal multilevel climate governance architecture. ASEAN climate policy continues to eschew intrusive supranational regulations and the collective target-setting that characterizes the EU's approach. Opportunities for regional collaboration on climate change and policy diffusion do exist; however, such mechanisms consists largely of "soft law" standards which provide member states with a range of optional policy instruments and guidelines that they may adopt at their discretion. Whether these opportunities lead to more ambitious regional action hinges on the willingness of national governments to seriously engage. So far, engagement with regional and transnational climate initiatives has been highly uneven and selective. Overall, MLG structures in ASEAN provide few access points for non-elite policy entrepreneurs, with existing national and regional institutions geared toward serving dominant interests. Where regional frameworks have occasionally offered potential policy solutions, they have been undermined by powerful national and sub-national elites.

Notably, regional climate action in ASEAN has been impeded by the lack of a powerful supranational policy entrepreneur, similar to the European Commission. ASEAN's focal actor on climate policy, the ASEAN Working Group on Climate Action (AWGCC), established in 2009, has few of the institutional prerogatives required to exercise an independent entrepreneurial function. Nevertheless, while our findings confirm the general claim that greater legalization and delegation of formal authority to supranational structures induces positive conditions for policymaking at the agenda-setting stage, it is important to not elevate Europe as necessarily an optimal model for MLG climate action. This echoes regional scholarship which is skeptical of claims of "European success" versus "Asian failure" (Katzenstein and Shiraishi 1997: 3). We find that, at least some of the time, deeper institutionalization in the EU belies a reality where success in opening and connecting agenda windows across Levels I and II must still contend with the challenge of opening the decision window at Level II and maintaining the integrity of the proposal through to the policy output stage. While this challenge is more acute in the

ASEAN setting, as we evidence, conflictual political economy dynamics surrounding ambitious climate action at Level II has also conspired in the EU to stagnate or even reverse policy agreed at Level I. In this way, comparing climate policy across the EU and ASEAN can provide helpful and useful comparisons.

Table 1 provides a summary of our EU and ASEAN case studies. We find that opportunity structures for policy entrepreneurs differ substantially, especially with regard to their ability to push for change in the politics stream. In the EU, the relatively dynamic interplay between different actors and levels has often conspired to couple streams and as a result made EU climate governance more ambitious (Jänicke and Quitzow 2017). As we explore in more detail in the subsequent chapter, at various points in time, this has resulted in non-incremental policy change, from the inception of the ETS to the EGD. Yet, at other times, the absence of agenda windows and policy entrepreneurs able to couple agenda-setting to decision-making has resulted in policy stagnation. Following the 2008 economic crisis, the hurdles for transformative policy change proved too high due to division both on Level I and Level II.

As Table 1 suggests, the hurdles for truly transformative policy change are even higher in ASEAN, reflected in a lack of substantial policy on the regional level. This is not explained solely by ASEAN's consensus-based and noninterventionist governance model. Rather, the primary obstacles to meaningful regional climate action are located at Level II, where effort to broaden problem definition and policy choice must contend with limited state capacity and/or politically dominant factions of capital that have become deeply enmeshed within state apparatuses. In contrast to the relatively open and dynamically contested policy processes in the EU, those in ASEAN are shown to be highly asymmetrical and dominated by small elite groups. The asymmetries of the politics stream mean that the policy problem is often deprioritized. While AMS' engagement with the UNFCCC has led to general acknowledgment of climate change, problem framing must contend with ASEAN elites' preferences for high-emissions economic growth models. Those same asymmetries further limit the development of the policy stream. Groups with the potential to advance more ambitious climate policy alternatives at domestic and regional scales of governance – reformers, civil society organizations, policy entrepreneurs – are either suppressed, or sidelined in policy forums where insider policy entrepreneurs hold sway.

5 Explaining Climate Policy Change in the EU: Multilevel Problems, Policies, and Politics

The EU has long sought to establish itself as a global climate leader. This aspiration has shaped its engagement with the UNFCCC regime and propelled

Table 1 Summary of EU and ASEAN case studies

Problem stream		EU Climate MSF	ASEAN Climate MSF
	Is the problem prioritized by elite policymakers?	Yes. Long-standing commitment to climate action on the international and regional levels, including efforts to promote climate leadership as a core identity issue.	No. Economic competitiveness consistently prioritized over climate action. External expectations on AMS traditionally also low.
	How is the problem framed?	Narrow problem definition (centered on emission cuts) giving way to more ambitious, broader framing under European Green Deal.	Framed primarily as a sustainable development issue, with emphasis on need for financial and technical assistance.
	How contested is the problem?	High value congruence, although deep divisions between climate "leaders and laggards."	Low value congruence, reflecting high political and socio-economic diversity among AMS.
	Who are the "problem brokers"?	Problem frames advanced by a plurality of actors (with Commission playing prominent role), albeit with bias toward technocratic framing.	Problem frames advanced by states/ elite centered on economic developmental economics, endorsed by powerful technocratic and business elites.

Policy stream		
Convergence on policy approach to problems?	Normative preference for legal/regulatory solutions (with exceptions).	Normative preference for facilitative and flexible solutions (with exceptions).
Technical capacity for policy development?	High technical and financial capacity (although with important national-level differences).	Low support for capacity-building at the regional level, significant capacity gaps between AMS.
Level of policy ambition?	Competitive "market" for potential regional-level policy solutions has driven higher ambition over time.	Little to no opportunity to advance mitigation policy proposals that would curb the power of dominant politico-economic interests.
Politics stream		
Who are the "power brokers?"	Comparatively dynamic political competition in the EU. However, elite interests enmeshed in EU structures and powerful states advantaged.	Political competition in ASEAN limited by powerful politico-economic coalitions in key states, which are tied to carbon-intensive industries in the region.
Ease of access to decision-makers?	Multiple access points for potential policy entrepreneurs. However, EU-level structures display limited democratic oversight.	Limited access points for non-elite policy entrepreneurs. ASEAN-level structures display no democratic oversight.

Table 1 (cont.)

	EU Climate MSF	ASEAN Climate MSF
Evidence of public opinion mobilization?	Public opinion data shows consistent and growing support for ambitious regional climate action, including costly climate mitigation action.	High general awareness, but least willing to bear the costs of climate change mitigation compared to publics in other regions of the world.
Evidence of pressure group impact?	Domestic pro-climate constituencies increasingly vocal and impactful on domestic climate discourse and policies.	Incipient mobilization of domestic pro-climate constituencies. However, activists are often subject to state repression, especially at the local level.

the development of a vast body of internal policies, laws, and regulatory instruments. At various points in time, the EU has implemented non-incremental policy changes, facilitated by a convergence of Kingdon's problem, policy, and politics streams across both Level I (supranational) and Level II (domestic). However, EU climate policy development has not advanced consistently over time and policy outcomes have not always matched European climate leadership aspirations. As the following analysis shows, European regional climate policy must be understood as the outcome of complex multi-level dynamics, with both external and internal drivers accounting for fluctuations in EU climate ambition over the past three decades.

This chapter is divided into two parts. We first identify key actors and institutions shaping developments in the EU's problem, policy, and politics stream. As Ackrill et al. (2013: 871) argue, analyzing the EU's MLG system through an MSF lens highlights the potential advantages of "what are normally considered to be pathologies of the EU system, such as institutional fluidity, jurisdictional overlap, endemic political conflict, policy entrepreneurship and varying time cycles." Thus, EU policy processes display conditions conducive to multilevel reinforcement as identified by Rietig (2020: 59) – level interdependence, regular cross-level interaction, recognition that problems on one level may require policy solutions on another level, and sufficient ambiguity to allow for venue shopping. However, while these conditions are necessary for multilevel reinforcement to occur, they are not sufficient. Both progressive and obstructionist policy entrepreneurs are able to strategically exploit the multiple access points offered by a "multilevel governance system [that] lacks transparency and comprises distinctive complexity and ambiguity" (Becker 2019: 148).

As our review of the three streams reveals, EU policy processes are characterized by path-dependencies and power structures that limit but do not preclude opportunities for pro-climate policy entrepreneurs to push for non-incremental change. In contrast to ASEAN, the EU has a powerful supranational bureaucracy – the European Commission – that exercises considerable influence on Level I, particularly in the problem and policy stream, where it is instrumental to opening Agenda Windows. However, in the politics stream, Level II political dynamics often determine which policy proposals make it over the finish line (Decision Window). Therefore, the second part of this chapter examines if, when, and how policy entrepreneurs within the EU have been able to exploit conditions for multilevel reinforcement. The European ETS and the EGD serve as case studies to evidence how policy entrepreneurs within the EU have successfully connected problems, policies, and politics across multiple levels. We contrast these experiences with the period following the global financial and economic crisis when EU climate policy development largely stagnated.

5.1 The Potential for Transformative Policy Change in the EU

This section provides an overview of the problem, policy, and politics streams in the EU, placing them in a multilevel context and identifying both opportunities and obstacles for climate-focused policy entrepreneurs. By outlining the actors and processes that keep these streams "flowing," it serves to provide context for the subsequent empirical investigation of when and how the three streams converge, creating windows of opportunity for transformative policy change. It also serves to facilitate comparison between policy processes in the EU and ASEAN. As we explore below, notwithstanding significant power hierarchies, relatively dynamic and open multilevel policy processes within the EU have enabled a diversity of actors to advance problem frames and potential policy solutions as well as influence opportunity structures in the politics stream.

5.1.1 Problem Stream

Climate action has long been a priority issue for the EU. Relatively narrow problem definitions – focused on emissions cuts – have gradually given way to broader and more urgent problem framings, emphasizing the need for transformational change. However, when it comes to delivering such change, techno-economic problem frames are still preferred. Within the parameters set by the European Council, the European Commission exercises substantial influence over which problems and problem frames are prioritized. In doing so, however, it is subject to a variety of pressures and events playing out at levels above and below.

The problem of climate change first entered European policy discussions in the early 1990s. Since then, joint climate policies have, at least in principle, enjoyed relatively consistent popular and political support, albeit with significant variations across member states. European institutions have repeatedly seized on the problem of climate change to reinforce their own legitimacy, advance the development of a joint foreign policy, and accelerate supranational identity building based on the "myth of a Green Europe" (Lenschow and Sprungk 2010).

Over time, dominant problem frames in the EU have changed, gradually transforming the issue of climate change from a narrow sectoral concern "into a high-politics, core-identity issue for the EU" (Torney 2015: 49). An important shift in problem framing occurred in the mid-2000s, when the European Commission pushed forward a more integrated approach to climate policy, mainstreaming it into other policy areas, notably energy (Skjærseth 2017). More recently, problem definitions have become even more expansive, with

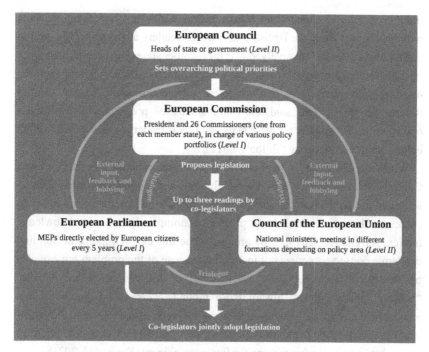

Figure 4 Decision-making in the EU (ordinary legislative procedure)

the 2019 EGD connecting the need for climate change mitigation and adaptation with a range of other ecological and societal challenges. Climate change is no longer framed simply as a "problem" but as an "emergency," requiring long-term transformations of key socio-technical systems rather than ad hoc emission cuts (Rankin 2019). Yet, notwithstanding the EU's high-level commitment to transformational change, current policies still reflect an understanding of sustainability that is steeped in techno-economic narratives of innovation and market rationalities (Olsson et al. 2021). The EU has also consistently sought to frame climate action as a means to increase competitiveness and green growth. However, in practice, internal market and economic growth imperatives have often competed with climate change and other environmental issues in the problem stream (Sadeleer 2014).

A plurality of actors within the EU is actively shaping problem framing. The EU's overall political and strategic directions are set at the high-level EU summits, when national and EU-level leaders convene as the European Council (see Figure 4). More than other EU institutions, the European Council serves primarily to reconcile national (Level II) interests. Through this venue, member states with higher bargaining power may be able to successfully push forward or resist particular problem frames. For example, the linking of the EU climate and energy agendas may not have been possible without support from the British and German leaders at the time (Bocquillon and Dobbels 2013). Conversely, powerful member states can also hit the brakes on more radical problem reframing, as exemplified by Germany's long-standing resistance to narratives that threaten techno-economic legacy frames (Oltermann 2021). That said, the rotating Council presidency also provides opportunities for smaller member states and their leaders to throw their weight behind particular agendas (Tallberg 2011).

While the European Council provides important impetus, it is the European Commission, operating at Level I, which holds a monopoly on formal agenda-setting in the policy process (see Figure 4). Key to the Commission's ability to exercise substantial influence on problem framing is its role as a knowledge broker. As Dreger (2014, 176) argues, gathering expert input and building epistemic consensus serves not just to make sense of the problem but can also be strategically employed to "technocratize" subsequent policy debates, "pull[ing] political actors toward those grounds where the Commission has a home field advantage." However, the Commission is also sensitive to the preferences of diverse interest groups (Coen et al. 2021) and it may choose to prioritize issues that have particular salience for European citizens (Koop et al. 2021).

While the agenda-setting powers of the EU's legislative bodies are more constrained, they too are able to wield a certain amount of influence on the problem stream. For example, the Environment Council, which brings together national environment ministers, plays a key role in preparing joint positions for UNFCCC negotiations, thus shaping the problem frames that the EU promotes at the global level. In turn, the European Parliament has increasingly explored informal avenues to raise the salience of certain issues through fostering public and/or elite support for policy action (Kreppel and Webb 2019). For example, the declaration of a "climate and environment emergency" by the Parliament in 2019, while largely symbolic, raised pressure on the EU's executive bodies to step up ambition (Rankin 2019).

Crucially, decisions which ultimately shape problem framing are not taken in isolation from societal realities and background developments of a technical and scientific nature. All EU institutions are responsive to what Kingdon (1995) conceptualizes as indicators, feedback, and focusing events. *Indicators* refer to new data and evidence that highlight a change in scale, scope, or nature of a particular problem. For example, IPCC assessment reports have regularly provided impetus for EU climate policy development. *Feedback* takes place when the evaluation of existing policies reveals implementation gaps, insufficiencies, or novel unexpected problems. This information may be gathered by policymakers and street-level bureaucrats, but also comes in the form of feedback from stakeholders and citizens and highlights important linkages between the problem and the politics stream. The recent youth climate strikes are arguably a particularly powerful form of feedback, since young people will be most affected by future climate change policies yet are not franchised across democratic political systems. Finally, *focusing* events refer to crises and emergencies that provide an extra push for problems to climb up the policy agenda (Kingdon 1995), such as heat waves or spikes in oil and gas prices.

Importantly, in MLG settings such as the EU, indicators, feedback, and focusing events play out at several levels, with EU policymakers responding to pressure from above as well as below. Having invested significant political capital into building a reputation as an international climate leader at Level I, the EU faces particular pressure to deliver on global climate treaties. At the same time, national (Level II) policy priorities, public opinion, and problem framing activities within member states, leveraged by interest groups with access to EU forums, influence which issues gain most salience in Brussels. At several points in recent years, this combination of top-down and bottom-up pressure has brought the problem of climate change to the fore and spurred the EU on to provide leadership in this area. However, "[i]t takes time, effort, mobilization of many actors, and the expenditure of political resources to keep an item prominent on the agenda" (Kingdon 1995) and, over the past three decades, climate change has been repeatedly crowded out by other policy challenges requiring urgent attention. In addition, a problem is unlikely to occupy policymakers' minds unless it can be coupled to concrete solutions, that is, if opportunities exist to open an Agenda Window. It is this second part of the agenda-setting process that we now turn to.

5.1.2 Policy Stream

The EU has historically demonstrated a normative preference for regulatory climate policy solutions. Compared to other regional organizations, the EU has a strong mandate and significant capacity for climate policy development and implementation. Although several EU institutions are involved in developing policy solutions, the Commission retains a monopoly of initiative. It plays a key role in gathering policy-relevant knowledge and collecting, assessing and selecting ideas that various policy entrepreneurs try to push onto the European level. Compared to ASEAN, the policy stream in the EU provides opportunities for a diversity of policy entrepreneurs to compete, although these opportunities are skewed toward a small circle of insiders that hold unparalleled access to EU policymakers.

Climate policy development in the EU has been driven by "a strong preference for regulatory instruments" (Jordan and Moore 2020: 57), combined with a limited selection of market-based instruments. To this end, climate policy-making has focused primarily on setting collective decadal targets, to be implemented through the EU's ETS and complementary effort sharing agreements, enshrining binding national targets for emissions reductions in non-ETS sectors. More recently, and in tandem with the transformation of dominant problem frames, the EU's climate policy toolbox has expanded. The EGD has shifted the focus decisively toward long-term, whole-economy planning in line with the goal of reaching collective climate neutrality by 2050. Other notable policy innovations include a Carbon Border Adjustment Mechanism (CBAM) which will impose a levy on some imports with a view to incentivizing emissions reduction efforts outside the EU and preventing "carbon leakage," that is, the transfer of production from the EU to countries with more lenient climate policies (Abnett and Twidale 2021).

In contrast to ASEAN, soft policy tools, such as voluntary targets and standards, remain relatively underutilized in the EU, not least because previous applications of such instruments increased doubt regarding their efficacy (Jordan and Moore 2020). Nevertheless, the EU continues to experiment with some soft tools aimed at catalyzing action from actors at other governance levels. An example is the EU Covenant of Mayors, a voluntary program which promotes action by local authorities to implement European sustainable energy policies (Domorenok 2019). The European Commission has also proposed a new voluntary standard, the European green bond standard, to encourage ambitious green investments that meet high-quality sustainability standards (Jessop 2021).

Among regional organizations, the EU is unique in being able to adopt and implement intrusive supranational policy tools. It has comparatively high administrative and technical capacities and EU institutions enjoy a broad mandate for regional policy development. The EU has been building up relevant expertise and capacities for decades, including in the highly technical area of emissions monitoring, reporting, and verification. However, it is important to note that – although its budget is large compared to that of ASEAN – the EU's financial autonomy remains very limited (Jordan and Moore 2020). Capacities also differ widely on the national level, with some member states able to go beyond what is required under EU climate legislation while others are reluctant to scale up ambition, for reasons that may include a lack of resources and implementation capacity (Sartor et al. 2019). This can lead to conflicting policy priorities at Level II.

Policy development is the result of input from actors located across EU institutions. While legislative proposals can only be adopted after review and adoption by both the European Parliament and the Council, the European Commission holds a formal monopoly of initiative. As such, it has the potential to provide significant cognitive, as well as entrepreneurial, leadership on climate change at Level I (Barnes 2010). However, it is important to note that the Commission is not monolithic. Although climate action has a dedicated Directorate-General (DG CLIMA), there are significant overlaps with the work of other DGs covering policy areas such as energy, environment, transport, industry, agriculture, and trade. Conflict between different DGs, due to competing interests, beliefs, and policy priorities, can dampen climate policy ambition (Rietig 2019).

Even when internally united, the Commission's entrepreneurial powers are not without limits. When developing policy proposals, it must consider the preferences of EU member states and parliamentarians. Both the Council and the European Parliament can make amendments to proposed policies and they may also invite the Commission to initiative legislation on a particular issue. The European Parliament also generates its own policy expertise through its influential environment (ENVI) committee. Committee rapporteurs, chosen to represent parliamentary positions in exchanges with the Council and the Commission, "play a central role in shaping policy outputs" (Burns 2019: 312). Thus, in practice, EU policymaking is a product of interinstitutional negotiations, usually facilitated by informal "trilogues" between the Commission, the Parliament, and the Council (see Figure 4), with ideas for policy solutions entering negotiations through various channels.

Ideas do not emerge out of thin air. As a comparatively small administration, the Commission is heavily dependent upon external expertise and input to develop policy proposals (Barnes 2010). Thus, its role in the policy stream is often that of an "entrepreneurial gatekeeper" – selecting, rejecting, or reshaping the ideas that float around in the "policy primeval soup" (Kingdon 1995), with a view to prizing open an Agenda Window. This soup is cooked up by a large and diverse community of specialists, which may work for nongovernmental and research organizations, industry associations and other interest groups, or public authorities at national or sub-national levels. The Commission and the legislative EU institutions regularly consult with different stakeholders to tap into this pool of expertise and to test the acceptability of policy proposals. While this makes for a relatively competitive landscape of ideas, there are important asymmetries in terms of which policy entrepreneurs gain privileged access to policymakers in Brussels. Navigating the EU bureaucracy requires time, knowledge, and resources, which are often not available to nongovernmental or not-for-profit climate policy entrepreneurs. In turn, the five largest fossil fuel companies alone have spent at least €250 million on EU lobbying over the past decade, spiking "at times when legislation is being drawn up" (Laville 2019). Fossil fuel–dependent industries and businesses, such as carmakers have also invested significant resources into lobbying EU institutions, at times successfully delaying stricter emissions regulations (Coen et al. 2021: 127). That said, the complex relationship between the EU and business is not simply governed by money and, ultimately, EU policymakers depend on a diversity of stakeholders for reliable information and expertise (Coen et al. 2021).

Like the problem stream, the EU policy stream is responsive to external developments on multiple levels of governance. As we explore further below, Level I ambitions vis-à-vis global treaties such as the Kyoto Protocol and the Paris Agreement have provided important impetus and the broad parameters for EU policy development. Conversely, some policy innovations have first been tested at Level II or promoted by non-state actors before being "uploaded" onto the European policy debate. The 2021 European Climate Law, which enshrines the EU's 2030 and 2050 mitigation targets in binding legislation, is a case in point, with prototype laws having first diffused at member-state level (Duwe and Evans 2020). Going forward, consideration of domestic context is likely to become increasingly important for EU policymakers. Sartor et al. (2019: 7) argue that, in order to coordinate a European-wide transition to net-zero, "the EU will need to significantly raise its capacity not simply as a legislator, but also a facilitator of national transitions that tackle in some cases quite different issues and priorities." This task is unlikely to be accomplished only via regulatory command-and-control measures, especially as the distributive consequences of the transition become more apparent. In this context, and as we explore in the next section, Level II politics are also likely to become increasingly consequential for efforts to open Decision Windows, that is, translating policy proposals into actionable outcomes.

5.1.3 Politics Stream

EU decision-making procedures are characterized by an "elaborate system of checks-and-balances" (Hix 2007: 147), albeit with constrained democratic oversight. Consensus-brokering is thus crucial to open Decision Windows that allow policy proposals to be turned into actionable policies. While veto-players cannot determine outcomes in the politics stream, they may be able to weaken ambition or negotiate substantial concessions. Domestic context is key to understanding negotiation dynamics in the EU's intergovernmental institutions, where coal-dependent Poland has emerged as the key challenger of higher climate ambition. Recent years have seen increasing polarization of the political debate, with growing public support for ambitious regional climate action on the one hand and growing concern over perceived EU regulatory "overreach" on the other. Generating broad societal buy-in on the domestic level will be important to support European transitional policies, as far-reaching and rapid interventions are now needed to ensure climate neutrality by 2050.

Just like problems and policies, politics in the EU unfold simultaneously on multiple levels. Given the elaborate EU decision-making procedures, consensus-building is a key prerequisite for any policy change. While no single member state, party, or interest group is able to prescribe policies in this "hyperconsensual" environment, the flipside is that concentrated minority interests may seek to block reform or ensure that adopted policies reflect the lowest common denominator (Hix 2007: 148). Another implication is that, in contrast to national-level policy cycles, Decision Windows are rarely suddenly flung open in the wake of an election or another significant political event. Rather, what is politically possible at any given moment is determined by a complex multilevel bargaining landscape, where the room for maneuver of policy entrepreneurs at the supranational level (Level I) may be severely constrained by developments within domestic (Level II) political systems.

Due to the salience of power dynamics at Level II, conflict between member states often dominates the headlines. With regard to climate change, divisions run principally (though not exclusively) between the "older" Western and Northern European member states and a handful of "newer" Eastern European member states, led by Poland. This makes it more challenging to adopt ambitious climate goals and recent efforts to do so have resulted in protracted negotiations.

Similarly to ASEAN, albeit to a lesser degree, economic development levels within the EU are not even and there are important differences in national context that determine how willing or able member states are to support ambitious climate action. Poland's energy-intensive economy, for example, remains heavily reliant on coal and this dependency is reinforced by a political economy that is characterized by strong links between the coal industry, the government, and miners' unions (Brauers and Oei 2020). Other Visegrád countries, notably Hungary, have often sided with Poland in EU climate debates, seeking to gain more financial support and other concessions (Dunai 2019). National context also shapes member states' bargaining positions on more specific issues. For example, Germany's decision to phase out nuclear energy – based on long-standing popular opposition – puts it at odds with a number of other EU members, including France, over the role of nuclear power in EU climate and energy policies (Barbière 2019).

Level II dynamics are dominant in this stream; however, politics also plays out in the EU's supranational institutions at Level I. The European Parliament has seen its formal legislative power increase during successive EU Treaty reforms and taken on a "more decisive role" in the EU politics stream (Becker 2019: 156: 156). While European elections are generally viewed as "second-order elections" (Hix and Marsh 2007), Braun (2021) finds that voters are increasingly motivated by genuinely European (Level I) rather than purely domestic (Level II) concerns. This may be particularly true for climate change, which is seen by a majority of European citizens as the most serious problem facing the EU (Eurobarometer 2021). Although its green ambition has fluctuated over time, the Parliament has often lived up to its reputation as an environmental champion (Burns 2019). In recent debates on EU climate targets, it has emerged as the most ambitious EU institution, pushing (unsuccessfully) for a 60 percent reduction in emissions by 2030 (European Parliament 2020) and greater stringency regarding target implementation (Taylor 2021).

However, recent European and national parliamentary elections have also pointed to growing polarization of political forces on both sides of the climate debate. While green pro-EU parties made substantial gains during the 2019 European elections, there was also rising support for right-wing populist parties, many of which deemphasize or oppose climate action (Waldholz 2019). Beyond electoral arenas, climate change politics in Europe have become more confrontational and fractious as pro-climate pressure groups and constituencies experiment with new social media platforms and direct action to voice discontent, from Extinction Rebellion's nonviolent civil disobedience to the youth-led climate strikes.

Growing polarization is evident not just in climate politics but also in the wider debate on European integration, which revolves around broader questions related to EU effectiveness and legitimacy, as well as entrenched power asymmetries between member states (Fabbrini 2015). This debate also puts the spotlight on the role of the European Commission, an institution that has historically sought to advance expert-based policymaking outside the politics stream, relying on technocratic rather than democratic legitimacy (Shapiro 2004: 345). In practice, the Commission increasingly operates at the interface of technocracy and politics. The election of the Commission President, for example, is no longer a barely noted administrative turnover but a highly political event. As we explore further below, the launch of the EGD was in many ways a political maneuver by the von der Leyen Commission, necessary to ensure support from the European Parliament as well as the European Council. Politics also plays out in the day-to-day work of the Commission, including in the form of internal power struggles and conflicts arising from overlapping responsibilities.

Politicization of climate change is likely to increase further as the EU and its member states have to contend with growing conflict over the distributive consequences of ambitious decarbonization policies. Absent robust government intervention to ameliorate social inequity, the costs, and benefits of climate policies are likely to be highly unevenly distributed across populations, fueling popular backlash (Abrams et al. 2020). Consequently, Level II politics can be expected to become increasingly consequential for climate policymaking at Level I (Aklin and Mildenberger 2020). While this means contending with powerful veto-players – from populist governments to domestic lobby groups to protest movements such as the "yellow vests" in France – there is also potential for vocal domestic pro-climate constituencies to push for more ambitious action. Encouragingly, in some member states that are generally considered climate laggards, growing public concern and grassroot mobilization on climate change appears to have facilitated some policy change. In Ireland, for example, where extensive lobbying by business and agricultural groups has long con-strained political preferences, recent spikes in public interest have "directly influenced" political decision-making on climate change, resulting in stronger legislation (O'Gorman 2020: 87). Even in Poland, there is indication that growing concern over climate change and local air quality has triggered con-crete, if modest, policy responses (Elkind and Bednarz 2020).

5.2 Climate Policymaking in the EU: Multilevel Reinforcement or Multilevel Stagnation?

In this section, we investigate how MLG dynamics – including interaction with the UNFCCC regime – may facilitate or hinder non-incremental policy change in the EU. The EU's unique policymaking system accommodates a diversity of policy entrepreneurs, albeit with significant disparities in terms of their power and access to decision-makers. MLG structures allow them to promote their preferred policy solutions on more than one level of governance, utilizing multiple policy venues and engaging in cross-level coalitions. Global level frameworks may be invoked to put pressure on policymakers, sometimes offering concrete parameters for policy development on the regional and/or national level. Policy entrepreneurs may also seek to promote policy innovations or best practices that first emerged on the national or local level. Under the right conditions, this can take the form of multilevel reinforcement, whereby the "open-ended and competitive governance structure of the EU" establishes "multiple and mutually-reinforcing opportunities for leadership" on environmental issues (Schreurs and Tiberghien 2007: 24).

However, positive reinforcement is not a foregone conclusion. In particular, policy entrepreneurs pushing for policy change on Level I may find it harder to gauge and promote political receptiveness toward new agendas – key to the opening of Decision Windows – since the direction of the politics stream is largely determined by Level II dynamics.

Below, we use three case studies to explore different outcomes of multiple stream dynamics within the robust MLG setting presented by the EU. To exemplify the potential for multilevel reinforcement, we drill down on two key climate policy initiatives: the EU's ETS and the EGD. We contrast these case studies with the experience of multilevel stagnation during the post-crisis period, when ambitious policy development in the EU largely stalled. In the interest of making the narrative more coherent, the case studies are presented in chronological order.

5.2.1 Multilevel Reinforcement and the EU-ETS: Putting Kyoto into Practice

The 1997 Kyoto Protocol proved a watershed moment for the EU, leading to the introduction of the first regional ETS. As one observer comments, "you might be able to draw a direct causal line between the design of the Kyoto Protocol and the shape of European climate and energy policy" (Interview EU Commission). The establishment of the ETS represented a genuinely non-incremental policy shift, given that the EU had originally been strongly opposed to emissions trading and the other flexibility measures that had been included in the Kyoto Protocol upon the insistence of the United States (Delbeke et al. 2015). Importantly, while the Kyoto Protocol introduced the possibility of carbon markets, it did not prescribe any specific policies. Thus, Kyoto participation alone cannot explain the EU's subsequent embrace of emissions trading as its flagship mechanism to deliver on international commitments. To explain this puzzle, developments across the EU problem, policy, and politics stream during the late 1990s and early 2000s must be understood within an increasingly dynamic multilevel setting.

In the problem stream, mounting scientific evidence on anthropogenic climate change and its impact (EEA 1999; IPCC 2001) invigorated the EU's commitment to deliver upon its international obligations. The EU had been particularly proactive during the Kyoto negotiations, notably committing to a higher emissions reduction target than other industrialized state parties, namely 8 percent between 2008 and 2012 compared to 1990 levels (Van Schaik and Schunz 2012). However, as yet, no internal measures and common policies existed within the EU to drive forward action toward this goal. Although GHG emissions across the EU fell throughout the early 1990s, this was mainly a result of structural changes in the economy of key member states, above all German reunification and the shift from coal to gas in the United Kingdom (Oberthür and Roche Kelly 2008). The slow pace of internal climate policy development and the lack of operational infrastructure to deliver on EU-wide GHG cuts resulted in a sizeable "credibility gap between international promises and domestic implementation" (Oberthür and Roche Kelly 2008: 39).

In the policy stream, the Commission established itself as the main entrepreneurial agent. This was partly a result of leadership changes within DG Environment, in particular the replacement of climate change unit leader Jørgen Henningsen with Jos Delbeke, who, as an economist, had a particular interest in carbon pricing and became a driving force behind the development of the ETS (Environmental Insights 2020). As it started exploring options for an EU-wide emissions trading scheme, the Commission also took developments on other governance levels into account. Given that some member states – namely the United Kingdom and Denmark – were already experimenting with emissions trading, the Commission's interest in an EU-level scheme was in part motivated by the desire to prevent the emergence of a fragmented European emissions trading landscape (Grubb et al. 2012). There was also an expectation that the establishment of the ETS would facilitate the development of a global carbon market, providing opportunities for linkages to other trading schemes and international initiatives (Braun 2009). Because the Commission itself did not possess much policy-relevant knowledge on emissions trading, it relied heavily on other actors for input (Skjærseth and Wettestad 2010). These included experts and consultants, environmental NGOs, business associations, and individual companies such as BP and Shell, which already had some experience with internal emissions trading (Braun 2009).

In the politics stream, the post-Kyoto moment presented an opportunity for the EU to demonstrate its capacity to act as a coherent and principled foreign policy actor. When the United States withdrew from Kyoto in 2001, the EU invested significant political capital in ensuring the survival of the Protocol. The EU's, ultimately successful, bid to rescue Kyoto "became not just an environmental goal but also a key aim of an emergent EU foreign policy by heightening European identification with the Kyoto Protocol" (Torney 2015: 49). Thus, the EU's political ambitions on the international stage provided additional impetus for the rapid development of EU-level policies. With regard to the design of such policies, emissions trading proved politically more palpable for member states than other options, notably plans for a carbon and energy tax previously advanced by the Commission.

The EU Emission Trading Directive was adopted by member states and the European Parliament in 2003 and the ETS became operational in 2005. During its initial two phases, it functioned in a very decentralized manner, granting significant autonomy to member states in terms of setting an overall cap for emissions. In addition, most allowances were freely allocated, rather than auctioned, and industry could use cheaper external credits from the Kyoto Protocol's flexibility mechanisms to comply (Lee 2014). These design flaws significantly weakened the environmental integrity and performance of the ETS. Nevertheless, the combination of EU measures taken during the first half of the 2000s likely had a substantial decreasing effect on GHG emissions in the second half of the decade (Dupont and Oberthür 2015), allowing the EU to overdeliver on its 8 percent emissions reductions commitments under the first Kyoto commitment period (EEA 2013). Such progress at EU level, however, stands in in stark contrast to the overall failure of Kyoto to galvanize concerted global action on GHG emissions which continued to increase at record rates until 2009 (IEEP 2020).

As this case illustrates, the inception of the ETS cannot be explained by focusing analysis on merely one level of governance. The European Commission emerged as a key entrepreneurial actor involved in orchestrating multilevel knowledge exchange and bringing together the problem, politics, and policy streams across various levels of governance. Crucially, the Commission knew that the Decision Window, dependent upon interest congruity at Level II, would remain shut for alternative policy instruments such as taxes and other fiscal levers, which – in contrast to emissions trading – required unanimous endorsement by member states. Political feasibility informed agenda choice from the start, with the Commission demonstrating not just analytical and managerial expertise but also the political acumen necessary to understand and navigate the needs and positions of different stakeholders in the decision-coupling phase (Mukherjee and Giest 2017: 11). The need to balance policy effectiveness and political feasibility explains many of the initial design flaws of the ETS, with the Commission seeking "to get the scheme underway in a form that elicited the most agreeable response by Member States and industry as well as environmental stakeholders while planning to delay some stricter restrictions to later development stages" (Mukherjee and Giest 2017: 14). Formulating ambitious policy alternatives that work in the political moment of the time remains a key challenge for European policy entrepreneurs. As we explore below, this was particular true during the aftermath of the Copenhagen climate conference and the economic crisis, when Decision Windows in the EU narrowed considerably.

5.2.2 Post-Crisis Europe: A Period of Multilevel Stagnation?

Following the establishment of the EU-ETS, most of the 2000s saw a further expansion of EU climate policies, including, most notably, the adoption of the first Climate and Energy Package, which enshrined three headline targets for

2020: (1) cutting GHG emissions by 20 percent from 1990 levels; (2) increasing the share of renewable energy sources to 20 percent; and (3) improving energy efficiency by 20 percent. These policies were consciously designed to enable positive "multi-level reinforcement, extending to the sub-national levels" (Jänicke and Quitzow 2017: 123). For example, the Commission set up the Covenant of Mayors in 2008 with the aim of facilitating the implementation of the Climate and Energy Package at city level and the subsequent upscaling of best practices (Kern 2018). However, toward the end of the 2000s, and following the financial and economic crisis, the window of opportunity for ambitious EU-level action closed, partly as a result of negatively reinforcing multilevel dynamics. In particular, the failure of the 2009 Copenhagen climate summit to deliver a Kyoto successor treaty meant that climate-progressive voices at Level I could no longer use the policy parameters set at UNFCCC level as a lever to bolster joint climate policymaking (Fischer and Geden 2015). At Level II, the post-2008 recession deepened divisions between member states on whether climate policies are necessary or detrimental for economic growth (Skovgaard 2014). Meanwhile, post-crisis public spending in the EU largely supported a fossil fuel driven recovery, also emboldening lobbying efforts by status quo industry interests at the domestic and European levels (Fernandez 2018). These dynamics all contributed to relative stagnation across the problem, politics, and policy streams.

At the turn of the decade, Europe started to feel the full fallout of the financial and economic crisis and the ensuing sovereign debt crisis. In the problem stream, concerns over unemployment, economic competitiveness, fiscal stability, and the future of European integration consequently crowded out climate-related concerns. This change in perception of problem urgency was also reflected in public surveys, with the percentage of Europeans seeing climate change as one of the most pressing global problems steadily declining post-2008 (Duijndam and van Beukering 2020).

Meanwhile, activities in the policy stream slowed as much of the Commission's previous entrepreneurial spirit had seeped away. Divisions between (and within) DG Energy and DG Climate Action could be exploited by less progressive factions of the Commission and also made those factions more accessible to business lobbying (Bürgin 2014; Fuchs and Feldhoff 2016). Moreover, in the absence of a new international treaty, climate-progressive voices within the Commission found it more difficult to develop a normative justification to significantly ramp up EU climate action. The Copenhagen summit had not just been a disappointment for the EU but also "a nadir for EU climate leadership" (Walker and Biedenkopf 2018: 36), with European negotiators finding themselves largely sidelined, while the United States,

China, and other emerging economies hammered out the nonbinding Copenhagen Accord. The EU had gone into the negotiations with a conditional offer to increase its own 2020 emission reduction target from 20 percent to 30 percent if other key countries committed to comparable efforts. Yet, this failed to convince the United States and others of the need for binding targets, dampening member states' enthusiasm for the EU's long-standing strategy of "leading by example" in global climate governance (Oberthür and Dupont 2011). Although the Commission made an attempt to unilaterally scale up ambition despite the Copenhagen outcome, it quickly gave in to pressure by member states and business groups, conceding that conditions for updating the EU's emission reduction target were "clearly not met" and citing "uncertainties" surrounding implementation of the Copenhagen Accord (European Commission 2010). Thus, whereas in earlier policy debates, international commitments had provided leverage to climate-progressive voices, "in the post-Copenhagen period, the changed circumstances were the most frequently cited argument for less ambitious targets" (Fischer and Geden 2015: 4–5).

In the politics stream, the EU now felt the "delayed political impact of the EU's eastern enlargement on climate policymaking" (Fischer and Geden 2015: 1). Poland emerged as a key veto player to more ambitious EU climate policy, backed by several other "new" member states. Yet, these states were not the only ones dragging their feet on more ambitious climate policies. As Bürgin (2014: 700) notes, a "lapse in leadership" from the traditionally climate-progressive member states encouraged laggards to be more assertive. This meant that slackening policy entrepreneurship in the agenda-coupling phase at Level I was accompanied by a lack of political entrepreneurship at Level II (Herweg et al. 2015) during the decision-coupling phase. The political landscape in the European Parliament also changed as growing EU skepticism manifested itself in the rise of nationalist-populist movements. The 2014 European parliament elections produced a "big bang" of populist, anti-EU parties whose agendas often reflected hostility toward climate action (Martín-Cubas et al. 2019).

While climate policy development in post-crisis Europe did not come to a halt altogether, it moved forward in a more incremental manner, with shrinking opportunities for policy entrepreneurs to push for path-departing change. In some cases, ambition had to be dropped significantly to push policy proposals through narrow Decision Windows. In other cases, Decision Windows remained shut. For example, although the EU began, for the first time, to explore in earnest the implications of a long-term climate target, namely, to reduce GHG emissions by 80–95 percent by 2050, the resulting low carbon roadmap for 2050, presented by the Commission in 2011, never received political endorsement from the Council due to a Polish veto. Even within the

Commission, as one observer notes, a serious commitment to a mid-century target was seen as "too ambitious" and "not helpful" by many (Interview NGO Stakeholder). A new Energy Efficiency Directive, adopted in 2012, was watered down in the legislation process, providing member states with a number of exclusions and exemptions and high levels of discretion in implementing the directive (Zygierewicz 2016). In 2013, planned EU regulations on reducing emissions from passenger cars were weakened after German chancellor Angela Merkel personally intervened in support of the car industry (Carrington 2013) and the credibility of EU regulatory efforts in the area was reduced even further in the aftermath of the 2015 Dieselgate scandal (Becker and Traufetter 2016).

In October 2014, after tense negotiations, EU leaders managed to agree on a second Climate and Energy Package, which introduced new headline targets for 2030: reducing GHG emissions by at least 40 percent from 1990 levels, increasing the share of renewable energy to 27 percent (binding only at EU level), and improving energy efficiency by 27 percent (indicative target). This was not an ambitious outcome as targets "did not go far beyond what would be reached with existing policies and, most importantly, were (partly) defined as 'non-binding'" (Fuchs and Feldhoff 2016: 58). According to Carey (2015: 7), the lack of nationally binding renewable energy targets was "partly a result of the perception that Brussels had been too intrusive in the case of the 2020 package," with member states wanting more control over their national energy strategies and energy mixes. It also reflected the result of intense lobbying by businesses, who had privileged access to high-level insiders, notably then-energy Commissioner Günther Oettinger (Fuchs and Feldhoff 2016; Fitch-Roy et al. 2018). Thus, as the EU went into the 2015 Paris negotiations, its internal policies largely reflected political achievability and concerns over cost-effectiveness rather than scientific urgency.

5.2.3 Multilevel Reinforcement and the European Green Deal: Putting Paris into Practice

Even after the adoption of the 2015 Paris Agreement, EU climate ambition was initially slow to pick up. Climate change did not enjoy the highest political priority in the face of other urgent challenges, from continued worries over economic growth to the refugee and migrant crisis, Brexit, and heightened security concerns over terrorism and a resurgent Russia. Policy action focused primarily on updating regulation in order to comply with the existing 2030 emissions reduction target. However, since 2018, multilevel reinforcement dynamics and a coming together of the problem, politics, and policy streams have moved climate change back up the policy agenda, culminating in the EGD.

Although many of its components build on existing EU policies and regulations, the EGD has been described as "revolutionary in concept" (Tsafos 2020). For the first time, it lays out an integrated, long-term plan for the transition toward a climate-neutral Europe by 2050, covering all sectors of the economy and also addressing issues such as waste, pollution, biodiversity, and sustainable food consumption. The EGD has been promoted by the Commission as "a new growth strategy" for Europe (von der Leyen 2019), and in the wake of the COVID-19 pandemic, it has become a key part of its economic recovery strategy (Dupont et al. 2020). The EGD also explicitly addresses the distributional consequences of decarbonization through a Just Transition Mechanism, designed to support fossil fuel–dependent regions in the transition toward net-zero and to get recalcitrant member states, notably Poland, on board. Another central element of the EGD is the European Climate Law, which turns the EU's 2050 climate neutrality objective into a legal obligation. The law also enshrines the EU's 2030 emissions reduction target, which has been raised from 40 percent to at least 55 percent to ensure consistency with the 2050 target and to demonstrate higher ambition in line with the Paris Agreement's five-yearly ratchet mechanism (Skjærseth 2021). Thus, as with previous EU climate policy outcomes, the emergence of the EGD cannot be fully understood without taking into account interactions between the international, regional, national, and subnational levels.

In the problem stream, the publication of the IPCC's Special Report on Global Warming of 1.5°C in 2018 was instrumental in raising awareness of the urgency and complexity of the climate change challenge (IPCC 2018). A Commission communication in November 2018 first introduced the vision of a climate-neutral Europe by 2050, explicitly citing the need to "respond to the recent IPCC report" and "lead the way worldwide" (European Commission 2018). This was also a direct response to the Paris Agreement which "really refocused the discussion on the mid-century goal and the long-term need for serious transformation" (Interview NGO Stakeholder). The sense of urgency was underscored by real-life events, as droughts and heat waves in Europe reached unprecedented levels, both in terms frequency and severity (Büntgen et al. 2021). Climate change now polled as the second most important problem facing the EU in Eurobarometer opinion surveys (Eurobarometer 2019). The declaration of a "climate and environmental emergency" by the European Parliament, following similar declarations by European cities, regions, communities, and states, confirmed that global warming was being recognized as a top concern (Rankin 2019).

In the policy stream, momentum for long-term, whole-economy strategic planning had been building up for some time. Important foundations for the

EGD and the European Climate Law had already been laid, for example, in the form of the 2018 Governance Regulations. As such, the EGD can be seen as the culmination of a longer process of climate policy transformation in the EU, "from narrow, separate climate and energy policy initiatives to broader coordinated packages aimed at achieving increasingly ambitious climate targets" (Skjærseth 2021: 26). The concept behind the European Climate Law also has a long history. Similar framework legislations had already been established in several member states before the Commission picked up the idea at a politically opportune moment. The international context also shaped opportunity structures for policy change. Both the Paris Agreement and the IPCC's 1.5°C report set the framework for a transition to net-zero by 2050, providing leverage for EU-level actors in support of increasing ambition (Skjærseth 2021). In contrast to the Copenhagen experience, the EU had played a key role in the negotiations at Paris and managed to achieve many of its policy objectives, meaning that leading European Commission officials were able to evoke a sense of ownership and stake in the implementation of what was seen as a "major win for Europe and its allies" (Cañete 2015).

In the politics stream, important changes at Level I included electoral gains for European green parties, as well as the confirmation of a new Commission under the leadership of Ursula von der Leyen, who immediately declared climate a "signature issue" (Farand 2019). Indeed, observers expressed surprise at "how aggressive the push by this new Commission has been" (Interview UNFCCC). Others concur that it "was a very hard-won battle" (Interview NGO Stakeholder). As Munta (2020) argues, it was partly due to the political contestation that accompanied von der Leyen's appointment that she made environmental sustainability a priority part of the Commission's work program. Her backroom nomination by the European Council left von der Leyen "in desperate need of a strong programmatic statement which would pacify the EP and simultaneously signal responsiveness to the European Council" (Munta 2020: 8), with green transitional policies promising the greatest overlap of interest.

Bottom-up pressure for more ambitious climate policies was also building up. 2019 saw unprecedented levels of public engagement on climate change, as evidenced by significantly increased media coverage (Pianta and Sisco 2020) as well as the success of the Fridays for Future strikes. This broader change in political mood at Level II also helped facilitate agreement on the EGD and stronger climate targets in the European Council. In December 2019, the Council endorsed the long-term goal of a climate-neutral EU by 2050. Although Poland secured an opt-out from implementing this objective, other EU leaders agreed to press ahead and two other hold outs – the Czech Republic and Hungary – dropped their resistance after securing guarantees on nuclear energy. In December 2020, after arduous negotiations, all member states agreed to the updated 2030 target (Mathiesen and Oroschakoff 2020). Thus, while Decision Windows were by no means wide open, leaders within countries that supported stronger climate goals were willing to act as political entrepreneurs (Herweg et al. 2015), facilitating decision-coupling through concessions and "carrots" such as the Just Transition Mechanism.

While the holistic approach of the EGD has been broadly welcomed by green campaigners, business associations, and other stakeholders, many of its specifics have invited fervent criticism. Above all, the 2030 emissions reduction target of at least 55 percent has been criticized as insufficiently ambitious. Going forward, a key question is how resilient commitment to the EGD will be in the wake of major crises such as the COVID-19 pandemic and the war in Ukraine. Both events provide strong reason and ample opportunity for more ambitious climate action; however, at the same time, they may divert political attention and resources away from the net-zero transition. As Dupont et al. (2020) argue, the political salience of the climate crisis enabled the Commission to provide continued policy entrepreneurship in the wake of the pandemic, joining forces with other stakeholders to turn the COVID-19 crisis into a window of opportunity and putting the EGD at the center of its recovery strategy. In contrast, the immediate threats to energy security resulting from the war in Ukraine have prompted some member states to resort to short-term solutions, such as an extension of coal use or the development of new infrastructure for liquefied natural gas (LNG), which risk strengthening fossil fuel dependency. It is too early to predict the long-term effects of COVID-19 and the Russian invasion of Ukraine on international, European, and national climate policymaking. Yet, the EGD clearly provides opportunities to link these and other global crises to the climate challenge. Moreover, the adoption of the European Climate Law makes it more difficult to deviate from agreed climate targets. As the Commission's Frans Timmermans emphasized, the fundamental idea behind the law is precisely to make sure that more immediate crises do not distract from these long-term objectives: "it allows you to focus on other things without losing track of what you need to do to reach climate neutrality" (qtd. in Rankin 2020).

5.3 Summary: Explaining Global Climate Policy Change in the EU

Relatively dynamic and open multilevel policy processes within the EU have enabled a diversity of actors to advance *problem* frames and potential *policy* solutions as well as influence opportunity structures in the *politics* stream.

Various policy and political entrepreneurs have stepped up at different points in time, from the European Commission to "green" member states and national leaders to sub-national, private, and civil society actors, with policy innovations often necessitated, inspired, or facilitated by developments on higher or lower governance levels, such as the adoption of international climate treaties or pressure from domestic pro-climate constituencies. Under the right conditions, these dynamics can spur policy innovation through multilevel reinforcement – creating a competitive landscape for climate leadership (Schreurs and Tibhergien 2007) and resulting in higher aggregate ambition (Rietig 2020).

Yet, dispersion of authority also poses challenges to transparency, legitimacy, and effectiveness, making the EU's internal decision-making process as well as its external system of representation uniquely complicated and cumbersome (Jordan et al. 2010). Thus, the EU is simultaneously "leaderless" and "leaderful" (Müller and Van Esch 2019). When policy windows open – that is, when Kingdon's problem, politics, and policy streams converge – a variety of individual actors, groups, and institutions can push forward policy proposals, enabling dynamic interactions, rapid diffusion of ideas, and possibly a "race to the top" in terms of ambition. However, MLG arrangements in the EU also provide ample space for well-resourced veto players and policy obstructors, seeking to prevent decision-coupling or obstructing the implementation of ambitious environmental policies at the national level (Laffan and O'Mahony 2008). Supranational (Level I) policy entrepreneurs, notably the Commission, must be attuned to the political moment, which is largely determined by power shifts within domestic political systems (Level II). Given the continued primacy of the domestic in supranational climate governance efforts (Aklin and Mildenberger 2020) – and the need to galvanize broad-based support for the transition to net-zero – a key question for the EU going forward is how to support domestic pro-climate constituencies whist ensuring "that no one is left behind" (von der Leyen 2019).

6 Explaining Climate Policy Change in the ASEAN: Multilevel Problems, Policies, and Politics

While climate policy and global leadership has been a prime concern for the EU, it is comparatively a less pressing issue for ASEAN. This is despite the fact that the ASEAN region is widely considered to be the one of the most vulnerable regions to the impacts of climate change (Huynh et al. 2014; ADB 2017). While contributing only 3 percent to global emissions, the ASEAN region recorded the largest emissions increase in the world between 1990 and 2010 (Raitzer et al. 2016). Under the Kyoto Protocol, AMS have had no mitigation responsibilities, with a fledgling regional commitment to mitigation only arriving with the Bangkok Resolution on ASEAN Environmental Cooperation in 2012. Subsequently, a regional climate governance framework has emerged in the form of a series of regulatory frameworks for the management of disasters, waste, and natural resources. These are further buttressed by energy-market integration initiatives such as the APG, and the AATHP aimed to curb forest fires. However, these frameworks do not contain emissions and decarbonization targets, raising doubts on their ability to deliver the climate ambition required by successive IPCC reports.

This chapter seeks to explain the limits of multilevel climate governance in ASEAN. Scholars have long argued that the "ASEAN Way" of regional integration sees member states adhering to principles of sovereignty, nonintervention, and consensus decision-making that limit the organization's ability to produce concrete outcomes on a range of issues. However, in adopting an MSF lens that focuses on actors and institutions shaping multilevel policy processes, we locate deeper "pathologies" within the ASEAN MLG system. We reveal a narrow and exclusivist climate policy process within ASEAN that centers around the interests and preferences of small cliques of political elites, leaving "outsider" policy entrepreneurs and climate activists with few access points at *both* Levels I and II. This picture stands in stark contrast to policy processes in the EU where a more open process allows both progressive and obstructionist forces to contest climate policy. Consequently, agenda windows can only open when problem frames and policy options are compatible with elite preferences. Nonetheless, the ASEAN climate framework, premised on an open-ended regulatory regionalism, facilitates a reasonable level of coordination for policy transfer and diffusion. However, the efficacy and ambition of such initiatives depend in large part on the willingness of national and sub-national governments to adopt and implement them. In the second half of this chapter, we therefore examine the implementation of the two most promising ASEAN initiatives – the AATHP and the APG – to illustrate how limitations in the domestic (Level II) *politics* stream undermine uptake and efficacy.

6.1 The Potential for Transformative Policy Change in ASEAN

In contrast to the relatively open and competitive policy processes within the EU, those in ASEAN are characterized by deep structural asymmetries in the *politics* stream. Here, powerful politico-business coalitions wield considerable influence over policymaking through regional, national, and sub-national governance institutions. Policy entrepreneurs and civil society actors, who often present a range of ambitious policy alternatives, are routinely excluded from or marginalized in regional decision-making that is largely monopolized by state actors. The asymmetries within the *politics* stream, in turn, limit the advancement of problem frames to those palatable to elite preferences for high-emissions economic growth. Similarly, these place severe constraints on who is able to forward policy alternatives, as entrepreneurs and activists with more ambitious proposals that run contrary to elite agendas are often sidelined. Nonetheless, a confluence of factors – engagements with UNFCCC after Copenhagen 2009, major climate disasters in the region, and pressure from global markets – have ensured a limited *agenda-coupling* where a modest but appropriate range of regional instruments (Level I) have been developed to promote "sustainable development." However, the latter suffer from a range of efficacy issues arising from problems in the domestic (Level II) *politics* stream where powerful elites continue to obstruct implementation, effectively crippling *decision-coupling*.

6.1.1 Problem Stream

ASEAN member states (AMS), who had no mitigation commitments under the Kyoto Protocol, have historically used the UNFCCC regime to defend a narrow interpretation of the common but differentiated responsibility (CBDR) principle. Beyond this, problem frames in ASEAN are limited by the monopoly state actors hold on regional and national agenda-setting that often exclude competing agendas from civil society. More recently, a confluence of factors – the bottom-up nature of the Paris Agreement; climate disasters in the region; and pressure from global markets – have initiated a modest shift in position from ASEAN elites. Climate change in ASEAN is now increasingly framed as a problem of sustainable development in order to legitimize economic growth priorities, while civil society attempts to reframe the climate crisis have been knocked back.

Since the Rio Earth Summit of 1992, where the Climate Change Convention was first agreed upon, governments of AMS have largely sought to deprioritize the problem of climate change. While ASEAN leaders did acknowledge the science of climate change and the necessity of global action, they did not accept that they should bear responsibility for it. From ASEAN's adoption of the Berlin Mandate at COP-1 to post-Paris Agreement, AMS have continued to prioritize economic development over climate goals. This was despite AMS largely accepting IPCC findings and recognizing their own vulnerability to climate change. AMS have attempted to use UNFCCC fora to "upload" their own policy preferences: to push for developed countries to (i) increase their own legally binding contributions to reflect historical emissions trajectories; and (ii) provide more generous financial and technical assistance to developing countries.

A significant stumbling block to climate negotiations between Annex-I and non-Annex parties has been different interpretations of the "common but differentiated responsibilities" (CBDR) principle. While the EU sees itself as taking on a leadership role, it also sees the responsibilities of developing countries as flexible and evolving. For AMS, on the other hand, the CBDR principle is taken to mean that the burden of mitigating climate change falls squarely on developed countries (Goron 2014: 109). This has led to gridlock in negotiations between more progressive climate leaders such as the EU, and non-Annex parties like the AMS, particularly in negotiating a successor to the Kyoto Protocol.

The framing of the policy problem in ASEAN (Level I) has remained largely limited because problem frames are almost always advanced by political and economic elites. ASEAN does indeed have some agenda-setting power through the chair, which rotates annually and alphabetically among members. Like the EU, rotating chairs can set the agenda with climate change formally on the ASEAN agenda since the Bangkok declaration of 2012. However, unlike the EU, the climate agenda has been limited because it is only state actors who get to set the agenda (see Figure 5). ASEAN's strategy of economic growth

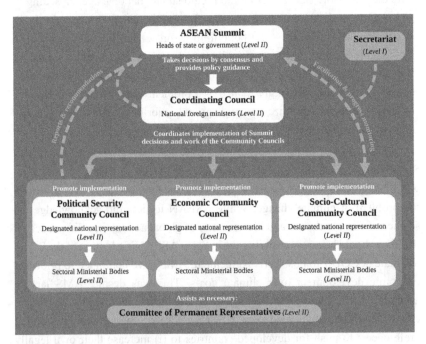

Figure 5 Decision-making in ASEAN

is described to be exclusionary in that it consistently excludes competing agendas, particularly in the space of human rights broadly defined, that contest entrenched models of economic development within individual states (Gerard 2014). ASEAN's attempts to engage with regional civil society has also been highly troubled – groups with agendas incompatible with ASEAN's preferred growth model, including certain environmental NGOs and activists, have been excluded from regional policy forums (Gerard 2014).

The ability of ASEAN to produce more substantive responses to the climate emergency and act as a regional "problem broker" is further undermined by low value congruence as, compared to the EU, the region straddles considerable economic and geographical diversity. AMS belong to different, and sometimes overlapping, groups in climate negotiations and have not managed to speak with a united voice. The China-led G77, where almost all AMS are members, is highly diverse and its position in climate negotiations has been described as a "lowest common denominator" approach that reflects this diversity (Goron 2014: 105). As a result, ASEAN lacks a "common narrative" in negotiations and regional joint statements on COPs do not stand out from G77 statements (Goron 2014: 108). Instead, AMS have participated in a range of diverse coalitions such as the Association of Small Island States (Singapore, somewhat inappropri-ately), the Like-Minded Developing Countries coalition (Indonesia, Malaysia, Philippines, Vietnam), the Least Developed Countries group (Cambodia, Myanmar, Laos until 2019), the Cartagena Dialogue for Progressive Action (Thailand, Indonesia), and the Coalition for Rainforest Nations (Indonesia, Laos, Malaysia, Thailand, Vietnam) (Goron 2014: 105). This diversity is further complicated by ASEAN's consensus model of decision-making where nothing is purportedly done without the consent of all ten members.

Despite these obstacles, developments leading up to the ratification of the Paris Agreement have facilitated modest progress in getting AMS to pay more substantial attention to the policy problem. This has worked in three interrelated ways. Firstly, the change of tact by climate leaders in engaging with the global south following the failures of Copenhagen in 2009 resulted in a more "bottom-up" Paris Agreement. The voluntary nature of Paris encouraged AMS to develop their own climate initiatives in return for funding and technical support. This facilitated opportunities for problem framing focused on both mitigation and adaptation, with climate initiatives such as the REDD+ program which seeks to minimize forest degradation being adopted in Vietnam and Indonesia (Hicks 2019; Pinandita 2020).

Second, adverse climate events have led to some acknowledgment among ASEAN leaders of the severity of the problem. In particular, the annual trans-boundary haze, caused by forest fires in Indonesia that spread to neighboring

Malaysia and Singapore, is a significant "focusing event" that regularly causes social and economic disruption. Coupled with the findings of the fourth and fifth IPCC assessment reports (2007 and 2014), this created an urgent sense among ASEAN leaders that climate change was indeed a threat to economic growth and livelihoods in the region.

Finally, as Southeast Asian economies become increasingly globalized, large domestic firms with strong political ties are facing pressures from international investment and consumer markets for private-sector climate disclosure and environmental and sustainability governance (ESG). This external pressure has prompted ASEAN governments to reframe the problem of environmental degradation as a problem of market access at a time when state development priorities are increasingly contingent on national economies being integrated into global markets.

These three developments contribute toward a very limited agenda-coupling where climate change is recognized but subjected to the priority of economic growth. Post-Paris, climate change in ASEAN (Level I) is framed primarily as a sustainable development issue with the emphasis on the need for financial and technical assistance. Within member states (Level II), there is some diversity on how the problem is framed. In more industrialized countries like Thailand, the problem of climate change is often connected to the vexing issue of domestic energy demand which is heavily reliant upon fossil fuels for its energy needs (IEA 2020). In Singapore, the problem is often framed as a market opportunity within the context of the country's post-industrial economic transition (National Climate Change Secretariat 2020). In economies dominated by extractive industries, such as Indonesia, the problem of climate change takes second place to the imperative of better management of natural resources to derive continuing economic benefit for politically influential agribusinesses (Wijaya et al. 2017). Overall, the framing of sustainable development post-Paris has allowed individual AMS to claim that dealing with climate change is compatible with economic growth priorities.

Within these limitations, NGOs and think tanks have nonetheless sought to reframe the policy problem at the national level (Level II) through public and closed-door advocacy with national governments and domestic industry. For instance, climate NGOs in Singapore have also sought to radically reframe the policy problem by challenging the government's narrow narrative on climate change. In 2020, the Activism in Crisis initiative, organized by a network of domestic NGOs including SG Climate Rally and Speak For Climate, sought to expand the problem frame of the climate crisis to include social justice and economic equity issues on the basis that "those who are already excluded by society and who contribute the least towards the climate crisis will be affected most by it" (Activism in Crisis 2020). However, such radical initiatives are rare in the region where climate agenda-setting at Level I continues to be monopolized by ASEAN elites and civil society actors are routinely subject to repression (see *politics stream* below). As a result, sustainable development has been seized upon in the ASEAN context as a means to legitimate economic growth in the context of environmental protection – to the detriment of problem frames which would compel a robust policy response in the form of stringent regulation of industrial development.

6.1.2 Policy Stream

The *policy* stream in ASEAN demonstrates a strong normative preference for facilitative and flexible policy solutions, with the ASEAN Agreement on Transboundary Haze Pollution (AATHP) a key exception. These policy solutions, promoted by insider policy entrepreneurs such as ASEAN bureaucrats and neoliberal economists, have promoted a reasonable range of natural resource management and energy-market integration solutions that can be adopted nationally and sub-nationally. However, unlike the EU, exclusivist regional and domestic participatory structures conspire to exclude or marginalize non-elite, or "outsider," policy entrepreneurs from forwarding more ambitious policy alternatives. Alongside dynamics in the *problem* stream, these present a limited *agenda coupling* where the agenda of climate change is articulated together with elite objectives and available policy solutions limited to a small compatible range.

The climate policy stream in ASEAN demonstrates a strong normative preference for facilitative and flexible solutions. ASEAN climate policy eschews intrusive supranational regulations and collective target-setting that characterize the EU approach. Instead, ASEAN climate governance consists largely of a series of "soft laws" that provide member states with a range of optional policy instruments and guidelines that they could adopt at home. Regional climate policy is largely organized under the purview of the ASEAN Working Group on Climate Change (AWGCC), formed in 2009, as a consultative body that would support coordination and collaboration between sectoral bodies of AMS.

Insider policy entrepreneurs at the AWGCC (Level I), often ASEAN bureaucrats and scientists, selectively "download" and develop a range of potential policy solutions from various global platforms that include natural resource management systems, coastal rehabilitation systems, biodiversity management and ecosystem rehabilitation tools, and peatland management systems (ASEAN 2018). At the same time, other regional platforms, particularly the AATHP, and the APG, organized under the purview of the ASEAN Economic Community

(AEC), also provide climate-relevant policy solutions for member states. AEC initiatives are often promoted by neoliberal technocrats or economists within ASEAN (Level I) forwarding market-integration agendas (Jones 2015: 6). With the exception of the Haze Agreement which is legally binding, these "downloaded" and adapted tools serve as a set of regulatory guidelines for member states to pursue their own environmental or energy market agendas, and provide platforms for knowledge-sharing, collaboration, and capacity-building. While they facilitate policy-transfer and capacity-building, current regional policy options, bereft of emissions and decarbonization targets, do not sufficiently drive the required climate ambition.

The search for collective climate solutions is a significant challenge for a region as diverse as ASEAN. AMS have significant differences in emissions profiles with vastly different national-level climate targets underpinned by different levels of state capacity and readiness. Singapore has emerged as a regional first mover when it comes to adaptation, in the form of desalination capacity, land reclamation, and sea walls to address rising sea levels. However, the country has not demonstrated any leadership at the regional level, though it strives for global market-leadership in green investments. There are perhaps limits to Singapore's ability to serve as a regional exemplar given that its emissions profile is relatively unique in the region with a large proportion attributable to the petrochemical industry, an almost nonexistent forestry sector, and – also unlike the rest of the region – an energy mix that is almost entirely reliant upon natural gas (96 percent).

In contrast, the region's largest emitter, Indonesia, must contend with a large proportion of its emissions originating in the forestry sector and its national climate plan rests heavily on reducing emissions from forest degradation. However, politics often has the final word on implementation. The Pakatan Harapan government in Malaysia (2018–2020) is indicative of the political contestation underlying climate policy, with the ruling coalition containing several rivalrous constituencies with some factions promoting progressive climate policies in opposition to others engaged in vigorous lobbying for the country's controversial palm oil industry (Varkkey 2019). Even among the progressive faction, progressive climate policies did not translate into concrete development of measurement, reporting, and verification (MRV) capacities as required by the Paris Agreement (Yeo 2018). When it comes to less developed AMS like Cambodia, Myanmar, and Laos, country-level capacity and resource constraints impede any move toward developing MRV capabilities, precluding the possibility of monitoring state or corporate emissions, let alone instigating an ETS (UNFCCC 2019). Unfortunately, pledges made in 2009 at the Copenhagen COP promising climate financing of $100 billion per year by 2020 to developing nations has failed to materialize (Carty et al. 2020).

ASEAN itself has no financing mechanism to assist poorer member states with the Asian Development Bank providing the main source of climate financing within the region, albeit with very limited funding for technical assistance (ADB 2021).

While the EU account demonstrates a competitive arena for potential regional-level policy solutions, albeit variable over time and within limits set by its commitment to technocratic institutionalism, exclusivist regional (Level I) and national (Level II) participatory structures in ASEAN place severe constraints on *who* is allowed to propose and advance policy alternatives. Relatively weaker levels of political accountability and participation in Southeast Asia have deep social and historical foundations. Cold War era economic development in pro-capitalist regimes saw the often-violent repression of independent civil society and the empowerment of regime-linked business interests (Rodan 2018: 44; Carroll 2020: 46). In the state-socialist regimes of Indochina, independent organizations were all but wiped out during this period of state reprisals. While subsequent processes of democratization produced divisions among elite factions and enabled a fragile reemergence of civil society in certain countries, the latter have never fully recovered from the often-brutal repression during the Cold War era. Organizational weakness within civil society has been further exacerbated by economic globalization and the neoliberal reforms of the post-Cold War era that increased the mobility and power of capital (Rodan 2018: 44).

These historical developments have produced illiberal regional and national institutions that are largely designed to satisfice the interests of key politico-business coalitions, while silencing or containing the policy preferences of non-elite actors. Indeed, not all policy entrepreneurs are made alike as regional and domestic policymakers often exclude or marginalize groups with agendas that run contrary to preferred models of economic growth. An excellent example of this phenomenon is Singapore's Emerging Stronger Taskforce which was formed to much fanfare in May 2020 to ensure that the country's post-COVID economic recovery has a strong focus on "green growth." Indicative of its function, the Taskforce is entirely composed of establishment and industry figures, including the Asia-head of Exxon Mobil and includes no civil society representation (Hicks 2020). While industry groups were invited to collectively propose policy solutions for recovery through the Alliances for Action initiative, citizen participation was relegated to a "dialogue series" for individuals called the Emerging Stronger Conversations (Tan 2020). At the regional level, while ASEAN has sought to buttress its legitimacy through greater engagement with civil society on policymaking following the Asian Financial Crisis, it has systematically excluded human rights and environment activists and groups whose agendas threaten dominant interests (Gerard 2014).

These exclusivist participatory structures work in practice to narrow the range of acceptable policy alternatives to the status quo, given that grassroots activists, NGOs, and think tanks have been key vectors for the diffusion of ambitious metagovernance norms in the region. The World Resources Institute (WRI) Indonesia, for example, is a well-resourced and prominent advocate for inserting the influential Science-Based Targets (SBTs) into the accountability structure and operational decisions of industry and government. Acting as a policy entrepreneur, WRI Indonesia has attempted to enhance transparency around carbon emissions by pushing for "data loops"[1] that could be used to ratchet up climate ambition on the part of government bodies and the private sector (Dagnet et al. 2019a).

Within the broader NGO sector, one climate activist from Singapore reports that they build strategy by "cherry-pick[ing] the best practices that are supported by science and implemented well in other contexts" (Interview with NGO stakeholder, Singapore). Other activists interviewed cited a broad range of "influences" on their advocacy strategies including IPCC reports and recommendations, more ambitious climate targets of other states, peer-reviewed climate science research, UK-based think tank Common Wealth, and even political ideologies like eco-socialism. However, their potential to exert influence is limited by the fact that many of these organizations lack a significant "social base" of support in society, and as such, find it difficult to gain political traction at a national level. Environmental NGOs in Indonesia, for example, have achieved important policy victories working closely with and advocating for local communities directly impacted by specific environmental disasters but this kind of activity generally fails to cut through at the national level.

For their part, private sector actors exposed to global consumer and investment markets have also begun to advance climate policy alternatives. Politically protected industries in the region have begun to advocate for carbon disclosure and transparency initiatives in response to the proliferation of ESG standards within global industries, as they seek to diversify their investment portfolios and expand beyond national borders (Al-Fadhat 2020). Public endorsement from national government has duly followed in some countries, such as Singapore. Elsewhere, the private sector has sought to reassure international investors by

---

[1] "Data loops" for climate action are based on the view that private firms sharing emissions data would inspire other firms to share similar data as well as setting more ambitious individual targets. Private sector climate ambition would, in turn, give governments greater confidence to enact stronger climate policies (see Dagnet et al. 2019b). Such an understanding is premised on what Ciplet and Roberts (2017: 150–151) term "neoliberal environmental governance" where "imperfect information" is taken to be one of several obstacles to environmental sustainability.

proactively filling the gap left by inconsistent state regulation, such as in Indonesia. The proliferation of private sector sustainability standards in the region, however, is haunted by the specter of greenwashing (Berliner and Prakash 2014). Private metagovernance norms such as ESG standards are often "downloaded" selectively and presented by business groups as "win-win" solutions which in reality resolve problems of market access but do little to address the underlying drivers of environmental degradation.

6.1.3 Politics Stream

The *politics* stream in ASEAN is dominated by small coteries of political elites with strong links to domestic business interests reliant on carbon-intensive economic growth models. Climate obstructors are thus deeply embedded within national, sub-national, and regional institutions that systematically exclude policy entrepreneurs and activists with competing agendas from formal policy processes. The primacy of powerful climate obstructors within multilevel policy processes means that *decision-coupling* is often crippled even when domestic capacities and appropriate policies are in the mix. Despite the promising emergence of an informal grassroots regime that monitors environmental degradation and advocates for more ambitious policy alternatives, climate activists currently lack significant broad-based support and continue to be suppressed and marginalized by coercive political regimes.

In contrast to the EU, the *politics* stream in ASEAN is highly asymmetrical. Climate politics is informed by the region's notoriously closed political economy dominated by powerful politico-business cliques with vested interests in sustaining carbon-intensive domestic industry. Reflecting the legacy of state corporatism in the region, climate policy obstructors are also often deeply enmeshed within the state apparatus, with national (Level II) and regional (Level I) institutions reflecting the extractive growth industrial preferences of this broad class of actors. Actors who fall outside this cabal have little access to political decision-making and highly circumscribed opportunities to put pressure on governments. It is possible that a nascent informal grassroots monitoring and advocacy sector could galvanize the kind of broad-based public support witnessed in Europe through the actions of Greta Thunberg and Extinction Rebellion. However, the challenge of building public support faces powerful structural constraints where regressive political forces will often exploit entrenched socio-economic inequity to frame climate change policies as contrary to the economic aspirations of the populace.

Despite some variation, the ASEAN region is heavily wedded to high-emission models of development centered on extractive industries dependent upon the use of fossil fuels (Gellert 2020; Hatcher 2020). Such "resource dependency" is tied to the interests of dominant elite groups (often politically connected conglomerates), as well as diverse constituents which rely on these industries for their livelihoods. As noted in the EU chapter, many businesses located in developed countries will also offshore their emissions to jurisdictions with laxer regulations. Taken together, global and local business elites have historically sought domestic market dominance through political patronage (with the notable exception of Singapore) resulting in favorable economic policies, large state subsidies, and the marginalization of competing societal interests, such as organized labor. In return, pliable political elites have bene-fitted from private sector largesse with donations being channeled to their partisan support bases. Despite the economic and political upheavals of recent decades, such politico-business coalitions have endured and remain deeply entrenched within the state apparatus. Key industries plugged into global supply chains, particularly forestry, mining, and fossil fuels (notably coal and petrol-eum products), have long enjoyed the patronage of political elites and this is set to continue as global and domestic demand for energy and resources climbs ever higher (IEA 2021).

These local structural realities often determine the outcome of political con-testation surrounding climate policy implementation, severely reducing oppor-tunities for *decision-coupling*, particularly at Level II. A key policy instrument in Indonesia to curb emissions through halting deforestation – the Moratorium on Primary Forests and Peatlands – serves as a useful illustration. While the Moratorium, in place since 2011, has had some positive impact on deforestation rates, it also displays significant limitations (Murdiyarso et al. 2011; Wijedasa et al. 2018) with illegal fires frequently breaking out in protected areas (Greenpeace 2017). The Peatlands Restoration Agency identifies local "interest groups" and provincial- and district-level governments as key sources of policy corruption, with the agribusiness lobby enjoying close ties to local government officials (Badan Restorasi Gambut 2016: 11–12). In forestry concession-heavy regions like Sumatra and Kalimantan, district heads are infamous for awarding forestry concessions to shell companies which are covertly owned by family members (Varkkey 2015). Concessions are then sold to commercial agribusi-nesses with the district head and their cronies pocketing the proceeds, rather than the district government. For their part, agribusinesses continue to be major funders of local politicians' election campaigns.

The energy sector in Indonesia is plagued by similar political issues that expand domestic demand for fossil fuels like coal. Indonesia is producing electricity in excess of domestic demand, and the present Widodo administration's industrial development plan is based on significant expansions in coal power (Arinaldo and Adiatma 2019: 4–5). Such deleterious environmental outcomes are a direct result of politico-business coalitions dominating the *politics* stream. Following substantial intra-elite friction in the lead up to the 2019 presidential election, President-elect Widodo formed a "grand coalition" of political allies and former rivals based around a national infrastructure project (Renaldi and Wires 2019; Wijaya and Nursamsu 2020). This national project would allow prominent political officeholders and their backers to economically benefit from infrastructural developments in return for their political support, as well as gaining popular legitimacy for the new government by delivering on GDP growth. The role of coal in driving this project is notable as key political figures and their allies hold coal-mining concessions or financial stakes in coal-fired power plants. For instance, Widodo's plan to move the Indonesian capital city from Jakarta to East Kalimantan will drive greater demand for electricity in the latter. Here, Widodo's supporters and former rivals will be well-placed to benefit as several hold land concessions in that area. A 2019 report by several environmental NGOs names notable individuals who have stakes in companies with land concessions, coal mines, and coal-fired plants in the region (Forest Watch Indonesia et al. 2019). These include the Coordinating Minister for Maritime Affairs and Investment Luhut Panjaitan, prominent oligarch Hashim Djojohadikusumo (brother of current Defense Minister and former presidential candidate Prabowo Subianto), head of Widodo's legal team Yusri Mahendra, and former Vice-President Jusuf Kalla (Forest Watch Indonesia et al. 2019, see also JATAM 2019). Predictably, the Indonesian government has continued to allow increases in coal production, with 2,000 new mining operations starting up in January 2018 alone.

Problems in the *politics* stream also manifest outside the domain of patronage politics. Some countries such as Singapore face a formidable dilemma when it comes to decarbonizing an economy which is wholly dependent upon the petrochemical industry for its basic function, with 96 percent of its energy use reliant on natural gas. Unsurprisingly it is the petrochemical industry, dominated by major fossil fuel corporations, which accounts for 75 percent of all industrial emissions and close to 45 percent of total emissions, making it, by far, the single largest contributor to GHG emissions in the country (Tan 2019). Contrary to its international commitments, the government is actively supporting the expansion of this industry. Indeed, the chair of the Inter-Ministerial Committee on Climate Change personally inaugurated refinery plant expansions for Total and ExxonMobil in December 2019 and March 2020, respectively (Prime Minister's Office Singapore 2019; Ng 2020).

Governments such as in Singapore face a hard reality, caught between the imperatives of decarbonization and continued economic development dependent upon heavy industry which cannot be easily transitioned to nonfossil fuels. Oil majors have been seen as essential developmental partners by the government since the 1970s (Ng 2012), fostering a "deep sense of reciprocity and goodwill" among both political elites and corporate boards (Interview with NGO stakeholder, Singapore). The economic salience of the petrochemical industry has only strengthened in recent years, as demand for non-oil exports decline and a socially fraught transition to a global financial hub which has undermined high-wage manufacturing jobs (Tan 2020). In this political context, the continued presence and expansion of oil and gas in the country is politically beneficial to the ruling party – the industry provides over 25,000 high-wage jobs. While social welfare provisions have expanded considerably (including wage-subsidies for low-income Singaporeans), since the ruling-Peoples Action Party's electoral setback in 2011, the party has yet to develop a more comprehensive plan to tackle long-term structural employment. This effectively means that on the climate mitigation front, the government's policy options are limited to technological innovations and market incentives.

At the same time, various opinion polls indicate that public support for ambitious climate policies has grown in the ASEAN and wider Asia Pacific region over the past decade (Kim 2011; Seah and Martinus 2021). However, they still lag behind similar polls for the EU (UNDP and University of Oxford 2021). Furthermore, none of the ASEAN states have fully declared a climate emergency, with only the Philippines and Singapore having declared "partial" emergencies in 2021 (Kurohi 2021; SunStar 2019). This raises the stakes for civil society in the region to challenge governments' lack of overall climate ambition as well as the structural factors that lead to environmental degradation. Correspondingly, we observe the emergence of informal grassroots regimes of civil society actors, operating largely at Level II, across the region that (i) monitor and report environmental degradation and climate policies, and (ii) advocate for policy alternatives either as policy entrepreneurs or political adversaries. While climate activism may appear unremarkable in Western liberal democracies, it has been an important driver for climate ambition and transparency, political accountability, and participatory policymaking. In contrast, climate activism in Southeast Asia is conducted within very limited political spaces where collective action can sometimes be criminalized or violently suppressed. Understandably, climate issues have rarely entered into mainstream electoral politics, though this could see some change in the coming years. In the lead up to the 2020 general elections in Singapore, two climate NGOs, SG Climate Rally and Speak for Climate, organized a climate scorecard for all political parties based on a number of metrics, the first of its kind (SG Climate Rally 2020). This was supplemented by a concerted campaign to get members of the public to individually lobby their political candidates to adopt more ambitious climate targets.

Despite significant inroads, the *politics* stream in the region continues to be characterized by significant power asymmetries. In countries such as Indonesia and Cambodia, environmental activists continue to be harassed by government officials and worse (Human Rights Watch 2020; Mongabay 2020). In Singapore, young climate strikers who individually posed for photos with placards were hauled up by the police for questioning over the country's strict illegal assembly laws (Han 2020). Most governments in the region continue to systematically exclude civil society with competing claims from climate and economic policy debates. The Singapore state, by far, demonstrates the most sophisticated mechanisms for controlling participation while excluding political contestation. NGOs are handpicked to attend closed-door consultations, with more radical or adversarial groups excluded. At these closed-door sessions, activists report that government representatives are more concerned with explaining the government's climate policies rather than seriously considering alternatives (Interviews with NGO stakeholders, Singapore). Publicly, government figures attempt to corral the support of climate NGOs in order to legitimize government policies.

A key weakness faced by many of these NGOs and think tanks in the region is that they lack significant social bases of support, and as such, find it difficult to gain political traction beyond their limited immediate constituencies such as local communities and domestic supporters. Environmental NGOs in Indonesia, for instance, often rely on working closely with and advocating for communities directly impacted by specific environmental disasters but enjoy little broad-based support beyond these. More concerningly, the Indonesian government's poor management of the COVID-19 pandemic has dampened societal mobilization against the unpopular Omnibus Bill that dismantles a number of environmental and labor protections (Jong 2020). Furthermore, mining companies and agribusinesses have leveraged a range of voluntary private standards to engage more with local communities which has had the impact of relocalizing contestation rather than allowing the scaling-up of advocacy efforts (Sinclair 2020). Variegated outcomes from these engagements tend to militate against the development of a broader domestic movement against extractive industries. While specific NGO initiatives in Singapore seek to mobilize more broad-based support for climate issues, these are still in their infancy with the SG Climate Rally, one of the largest groups, only starting in mid-2019. Movements that frame the climate crisis around principles of distribution and justice are starting to take shape and will require significantly more broad-based support to start to properly challenge powerful entrenched interests.

6.2 Multilevel Climate Governance in ASEAN: Asymmetric Politics and the Problem of Domestic Implementation

The above section demonstrates both opportunities and obstacles to multilevel climate governance arrangements in ASEAN. Despite variations in national circumstances and the absence of a "pooling sovereignty" model, multilevel dynamics have indeed produced modest opportunities for greater climate ambition. The bottom-up nature of the Paris Agreement and AMS' individual interaction with the UNFCCC, amidst regional environmental vulnerabilities and the internationalization of domestic conglomerates, have facilitated a modestly limited *agenda-coupling*. However, obstacles situated in the highly asymmetrical *politics* stream dominated by powerful politico-business alliances not only prevent *decision-coupling*, but indirectly limit *agenda-coupling* by narrowing the scope of the policy problem and restricting the range of admissible policy alternatives. At the same time, ASEAN's regulatory regionalism framework does indeed provide further opportunities for regional collaboration on climate change and policy diffusion. However, whether these opportunities lead to more ambitious regional climate action hinges on the willingness and capacities of national governments to engage and adopt available solutions.

The remainder of this section will examine the implementation of two key multilevel initiatives in the region – the AATHP and the APG. This is done to demonstrate how power asymmetries in the domestic (Level II) *politics* stream obstruct *decision-coupling* by undermining or limiting the efficacy of promising regional (Level I) climate policy instruments.

6.2.1 ASEAN Agreement on Transboundary Haze Pollution

The AATHP was created in 2002, largely driven by the governments of Singapore and Malaysia, in response to the transboundary haze crisis caused by forest fires for clearing land for commercial palm oil and timber in Indonesia. Unlike most ASEAN agreements, the AATHP is legally binding, and predates the Paris Agreement by over a decade. By December 2003, over half of AMS had ratified the agreement, with Indonesia holding out the longest but finally ratifying in 2014. The agreement is multilevel in nature and is integrated into both regional and Indonesian domestic forestry and peatland management initiatives. It is further complemented by Indonesia's Moratorium on Primary Forests and Peatlands that has been in place since 2011 as a condition for REDD+ funding. The AATHP effectively rescales the governance of peatlands and forest fires in

Indonesia by creating a regional task force of technical experts to provide regulatory standards for national and sub-national governance actors to monitor and intervene in forest fires. While not primarily designed to target GHG emissions from its inception, the agreement has sufficient teeth to tackle forest fires on a transnational scale by allowing regional actors access to local points of regulation, with domestic capacities further buttressed by international pressure and aid. On paper, such arrangements create the potential for multilevel reinforcement for curbing GHG emissions through preventing seasonal forest fires.

As with the national moratorium (see Section 6.1), the AATHP has been regularly undermined by local elites in cahoots with agribusinesses (Hameiri and Jones 2013). District heads in Indonesia have been known to deny regional experts access to politically protected plantations, while also intentionally underfunding sub-national institutions that are part of the regional fire control architecture (Hameiri and Jones 2013: 471). Sub-national political elites have also consistently resisted efforts to rescale haze prevention to the regional level by insisting on the deployment of local firefighting resources. This often meant that fires were not tackled until the burning was completed (Hameiri and Jones 2013). These problems are not just consigned to the sub-national levels in Indonesia. As kickbacks from agribusinesses funnel upward to national politicians, public discourse in Indonesia is increasingly framed in terms of protecting national development and "national interests." In 2019, amidst growing controversy, the Widodo administration instructed agribusinesses not to publicly share information on their palm oil concessions, with Greenpeace accusing the Indonesian government of "actively blocking" attempts to reform the troubled palm oil sector (Greenpeace 2019). Such moves acutely reduce transparency and public accountability in forest fire governance, further undermining regional attempts of monitor and tackle forest fires.

The problems in Indonesia's domestic *politics* stream that undermine regional climate governance have deep historical roots. The New Order era from 1965 saw the regime open forests to logging companies and agribusinesses where control over forests became increasingly concentrated in the hands of oligarchs and state-owned forestry concessions (Gellert 2015: 78–79). While the global demand for palm oil has been often cited as a cause for forest degradation (Indonesia is currently the largest global supplier), the induced domestic demand for palm oil is equally significant. This happened through deliberate government policies from the 1970s that substituted cheaper forms of oil (particularly coconut oil) with palm oil (Gaskell 2015: 30–37). The World Bank also actively promoted palm oil as a "development crop" in Indonesia since 1965 (Gellert 2015: 80). Neoliberal market reform and political devolution following the fall of the New Order accelerated the expansion of extractive forestry. Political decentralization led to a further proliferation of patronage networks, with Singapore- and Malaysia-based palm oil companies, backed by their respective governments, getting in on the act as palm oil production became increasingly regionalized (Varkkey 2015).

Consequently, regional haze-control initiatives have ended up protecting agribusinesses due to close connections between domestic political elites and agribusinesses, and the ensuing dominance of these networks within the *politics* streams of key AMS. Haze prevention efforts end up being focused on small-holders and local communities who lack powerful political backers (Hameiri and Jones 2013: 471–472). For example, the now-discontinued "Adopt-a-District" programs led by the governments of Singapore and Malaysia targeted small-scale slash-and-burn farmers when commercial plantations account for 80 percent of the haze (Varkkey 2015: 195–199, 215). This allowed these two respective governments to appease public outrage against the haze by targeting "elements that did not affect their plantation business interests in Indonesia" (Varkkey 2015: 202).

Non-elite policy entrepreneurs, particularly conservation NGOs, have none-theless emerged to alleviate environmental outcomes, seeking both collabora-tive and adversarial engagement with agribusinesses. These engagements largely revolve around agribusiness-led forums on the implementations of transnational voluntary sustainability standards in the palm oil sector, notably the Roundtable on Sustainable Palm Oil (RSPO). However, the ability of NGOs to impact outcomes within such forums is limited by the cooption of collabora-tive groups for the legitimation of business agendas and the exclusion of more adversarial groups (Ruysschaert and Salles 2016). With little or no access to domestic and regional policy forums, adversarial groups have responded by publicly highlighting the environmental and human rights abuses of individual companies in order to elicit public responses from multinational corporations that regularly use palm oil for their products. One such report by Friends of the Earth (2022) has led to large multinationals such as Nestle and Danone cutting ties with palm oil producers accused of environmental harm and human rights abuses (Sirtori-Cortina et al. 2022). Disconcertingly, palm oil imports are growing in markets (particularly in China and India) that do not demand conformity to sustainable standards (Dauvergne 2018), thus significantly redu-cing the leverage of non-elite policy entrepreneurs.

6.2.2 ASEAN Power Grid

The idea of a regional power grid to meet regional energy needs was first floated in 1998, with a memorandum of understanding signed in 2012. By the end of

2015, the APG became a fully functional part of the broader AEC that sought to build an integrated regional market through eliminating domestic barriers to trade and investment. The APG follows the "2+X" approach of the AEC where two member states can proceed with regulatory reforms without waiting for others to be similarly prepared. The APG provides a good example of ASEAN's "open" regulatory regionalism in that regional institutions provide an overarching regulatory framework that national governments can adopt to pursue their own energy market priorities. As with the AATHP, the APG is not primarily designed as a climate governance instrument. However, a regional grid provides opportunities to facilitate a greater share of renewable energy sources into domestic energy markets, while allowing more developed countries (like Singapore) with acute spatial limitations in adopting renewables. The APG would, thus, facilitate energy trading between countries with ample renewable sources to countries with lower access to renewables, which would in turn, foster greater investments in renewable sources in less developed countries (Ahmed et al. 2017a).

While the APG has taken off, it suffers from two critical limitations. Firstly, energy grid projects have suffered from low rates of adoption. APG projects have been largely limited to trading between Thailand and its less developed neighbors (Cambodia, Myanmar, and Laos), with some limited involvement between Indonesia (the region's largest electricity consumer by far), Singapore, the Philippines, and Malaysia (Ahmed et al. 2017a). Similarly, multiple reports have also indicated that the APG appears a set of "bilateral conduits" rather than an integrated regional network (Jones 2015: 15; Ahmed et al. 2017b). At Level II, policy entrepreneurs operating there reported considerable national government circumspection toward the prospects of furthering APG connections and "citing national energy security" as a reason (Interview with NGO stakeholders, Singapore and Indonesia).

At the same time, insider policy entrepreneurs operating largely at Level I, such as the Economic Research Institute for ASEAN and East Asia (ERIA) and the Energy Studies Institute (ESI) at the National University of Singapore, have stepped in to support ASEAN in building a more integrated regional energy market. Keen to avoid the perils of taking sides in domestic politics, these insider policy entrepreneurs have largely focused on the technical limitations of the APG (ERIA 2015; Andrews-Speed 2016). Notwithstanding these limitations, the ERIA and the ESI have proved to be a useful focal point for the diffusion of metagovernance norms to govern integrated energy markets. For instance, the ESI (at Level I) has actively promoted the pan-European Nord Pool power exchange model, successfully replicated in India and Southern Africa, as a viable model to facilitate greater integration around the APG

(AEMI 2015; Andrews-Speed 2016). Similarly, the ERIA have also promoted the use of Distributed Energy Systems (DES) as a technical solution that would support both ASEAN (Level I) and individual member states (Level II) in increasing their share of renewables through lowering capital costs (ERIA 2018).

Despite promising policy entrepreneurship at Level I, the APG is undermined by powerful domestic politico-business coalitions in the *politics* stream at Level II that obstruct *decision-coupling*. APG agreements would necessitate the removal of long-standing domestic fuel subsidies to energy consumers (Victor 2009). This has, in turn, prompted significant resistance from state-owned energy firms and politically influential energy-consuming industries (particularly in Indonesia and Malaysia) that owe their dominance to these long-standing subsidies (Jones 2015: 16). The removal of subsidies would also affect the incomes of workers and low-income groups in the region, causing concern to political elites as the distributive impacts of fuel subsidies are key to the durability of many political regimes in the region (Jones 2015). Beyond domestic fuel subsidies, transborder energy projects can be unattractive for local elites in areas dominated by patronage politics. Karim (2019: 1,564), for instance, documents how APG projects between Sarawak (Malaysia) and West Kalimantan (Indonesia) were slowed down by local political elites in the latter because (cheaper) energy imports from Malaysia would undercut lucrative local energy projects. Development of the latter, while eschewing cheaper APG sources, allows local political elites in West Kalimantan to access kickbacks from economic elites in return for approving energy projects.

Secondly, the APG is falling short on the renewables front. While the APG has targeted a 23 percent share of renewables by 2025, this would involve the region doubling its current share of renewable energy sources which does not appear on the cards, even for ASEAN (The ASEAN Post 2019). Rather, APG participants are prioritizing the affordability and availability of fuel types rather than environmental sustainability (Ahmed et al. 2017b). This has led to the overwhelming use of fossil fuels within the grid despite the availability of renewable sources in the region (Ahmed et al. 2017b).

Asymmetries within the domestic *politics* stream also explain limitations to the uptake of renewables in the broader region. The coal industry in Indonesia, for example, receives fifteen unique government subsidies, only seven of which have been quantified by analysts (Clark et al. 2020). In 2015 alone, these seven subsidies were estimated to be around US$ 946 million, while renewables only received a cumulative subsidy of US$ 179 between 2010 and 2015 (Clark et al. 2020: 8). Further complicating matters is the position of the state-owned electricity provider Perusahaan Listrik Negara (PLN) within these patronage networks.

Energy regulations in the country emphasize domestic self-sufficiency, which the PLN leads by working closely with domestic fossil fuel companies that are either state-owned or part of oligarchic networks (Bosnia 2018; Guild 2020). Simultaneously, the PLN effectively prices out renewable energy producers through convoluted tariff negotiations and stipulations on local content that preclude the benefits of cheap imported renewable technology from China, among others (Wicaksono 2015; Guild 2019). The PLN has also been reported to have flatly refused to buy solar energy from fossil fuel producers looking to transition away from coal (Maulia 2021). Furthermore, the tariffs levied by the PLN are often in contradiction with the policies of the Ministry of Energy and Mineral Resources, which, similar to ASEAN, is aiming for a 23 percent renewable energy share by 2025 (Agustinus 2016; Walton 2019). Notwithstanding this, the Ministry itself has previously imposed its own tariffs on renewable energy, effectively disincentivizing investments into renewables (The Jakarta Post 2017).

Due to these entrenched elite interests at Level II, as well as the limited political space afforded to competing demands at both levels, policy entrepreneurs have found it difficult to actively advocate for the expansion of renewable energy. Even in countries like Singapore where the government has been more open to sourcing for renewable sources, policy entrepreneurs there (Level II) have reported that their advocacy efforts have been thwarted by domestic politics in neighboring countries. Indonesia and Malaysia, for instance, have recently curbed renewable energy exports to Singapore in order to boost their own faltering renewable targets (Free Malaysia Today 2022; Vietnam Plus 2022). This limits the impacts of advocacy for the adoption of renewable sources with activists in Singapore stating that any cross-border collaboration on renewables will have to accommodate the interests of elite groups on both sides (Interview with NGO stakeholder, Singapore). At the same time, international NGOs such as Greenpeace have also sought to put global pressure on ASEAN governments to phase out coal and increase their renewable shares through a highly critical Power Sector Scorecard for individual AMS (Greenpeace 2020). This is further complemented by their direct advocacy toward investment banks in East and Southeast Asia to halt further investments in coal power plants (Greenpeace 2021). Despite these pressures, any optimism toward the APG reaching its renewables target should be tempered with caution.

6.3 Summary: Explaining Global Climate Policy Change in the ASEAN

In contrast to the relatively open and dynamically contested policy processes in the EU, those in ASEAN are shown to be highly asymmetrical and dominated by small elite groups. These asymmetries originate from the domestic *politics* stream where domestic politico-business coalitions, which have historically benefitted from carbon-intensive growth, wield considerable political influence over domestic and regional institutions. The asymmetries of the *politics* stream mean that the policy *problem* is often deprioritized. While AMS' engagement with the UNFCCC on the road to the Paris Agreement has led to a better acknowledgment of the problem of climate change, it is often articulated together with and tempered by ASEAN elites' preferences for high-emissions economic growth models. These same asymmetries further limit the development of the *policy* stream. Groups with the potential to advance more ambitious climate policy alternatives at domestic and regional scales of governance – reformers, civil society organizations, policy entrepreneurs – are either suppressed, or sidelined in policy forums where they exist.

These restrictive policy dynamics, coupled with an ASEAN regional structure centered on regulatory regionalism, offer extremely limited opportunities for civil society and policy entrepreneurs to exploit potential windows for transformative policy change. The Paris Agreement, with its pledge-and-review structure, has indeed led to more constructive engagement between climate leaders and AMS. This constructive engagement has led to the downloading of global policy solutions to the regional level where the ASEAN regional framework, in turn, creates opportunities for regional collaboration and policy diffusion. These dynamics demonstrate a limited form of *agenda-coupling* where climate change is recognized as a sustainable development issue that is to be addressed by a range of modest policy options without the necessary emissions and decarbonization targets.

However, the domination of domestic policy processes by powerful politico-business coalitions means that *decision-coupling* is severely hampered. As this account has shown, regional frameworks are either unevenly adopted or undermined by these interests during implementation. MLG structures in ASEAN provide few access points for non-elite policy entrepreneurs, with domestic and regional institutions largely geared toward serving or highly accommodative toward dominant elite interests. The structural asymmetries highlighted in the ASEAN case suggest that they cannot simply be overcome by agile policy entrepreneurship alone. Transformative policy change hinges on the extent to which pro-climate constituencies are able to scale up their efforts and mobilize broad-based support for more ambitious climate targets and policy instruments. Global climate support for the global south needs to urgently go beyond extending climate financing and technical assistance to governments and the private sector, to actively supporting and resourcing the scaling-up of pro-climate constituencies.

7 Conclusion: What Next for Climate Policy in the EU and ASEAN?

As we approach the ten-year anniversary of the Paris Agreement, it remains unclear whether the global climate regime will deliver the rapid and far-reaching measures required to prevent catastrophic global warming. The symbolically important 1.5 degrees target is quickly slipping out of reach, as global GHG emissions will have to peak before 2025 to achieve even the less ambitious 2 degrees target (IPCC 2022). We are, in other words, at a decisive crossroads for climate policy. What is not in doubt is the extraordinary groundswell of policy activity since the Paris meeting in 2015. Much of this activity has occurred within regional organizations, which have emerged as key sites of "agency between the nation-state and global institutions," well-placed to mediate climate norms from the international level to regional-specific political, institutional, and social realities (Börzel and van Hüllen 2015: 3).

This Element provides novel empirical insights into the opportunities and obstacles for climate policy development in two key regional organizations, the EU and ASEAN. More specifically, it inquires into the conditions under which climate-progressive policy entrepreneurs in these settings are able to successfully exploit interlinkages between different levels of governance and secure transformative policy change. By combining an MLG perspective with John Kingdon's MSF, we develop a conceptual framework which accommodates a concern for agency while also taking seriously the structural conditions that shape global-to-regional-to-local policy delivery. The Element contributes to comparative research on the EU and ASEAN, which, to date, has focused mostly on variation in broader integration dynamics (Acharya 2013). It also offers a basis for refining the MSF concept, which has found limited application outside of the "Global North" and in the context of an MLG reality.

Explaining transformative policy changes remains one of the hardest tasks in the policy sciences (Baumgartner and Jones 2009). Our findings confirm the general utility of the MSF for explaining such outcomes within MLG settings defined by institutional ambiguity and multiple venues. As evidenced especially in the EGD, policy entrepreneurs – in this case, above all the EU Commission – who are able to exploit cross-level interdependency, coalition building, and issue-linkage can spur policy innovation through "multi-level reinforcement," creating a competitive ecosystem for climate leadership conducive to higher aggregate ambition (Schreurs and Tibhergien 2007). However, such transformative policy outcomes are highly contingent on the three streams converging to allow policy equilibria to shift, with the room for maneuver at the regional level often constrained by domestic veto players.

While our EU case study suggests that, at crucial moments, synchronous opening of windows of opportunity from "above" and "below" have been decisive, our ASEAN case study serves as a counterfactual. In the absence of a European Commission-type entrepreneur, alongside auxiliary actors, the task of problem framing and policy development has fallen upon insider policy entrepreneurs (scientists and technocrats) within the ASEAN machinery. As we evidence, the resulting promotion of open-ended technical and market-integration instruments have offered little hope of transformative policy change. Where MLG structures in ASEAN have provided fleeting but meaningful opportunities for multilevel reinforcement, progressive action has almost invariably been thwarted at lower levels. The AATHP and the APG show how promising regional initiatives run aground on the rocks of domestic veto players.

Such MLG implementation challenges go far beyond questions of technical efficacy, implicating local configurations of power and political economy. Whether the problem, policy, or politics stream, each contains limits on who is allowed to contest and promote policy alternatives and on what terms. It is noteworthy that the variation in successful coupling has, on occasion, followed similar patterns across the ASEAN and EU. In both the EU and ASEAN, we find that MLG dynamics have facilitated agenda-coupling, albeit to varying extents. In both cases, transformative problem frames and policy alternatives have been available, though relatively more restricted in the ASEAN case. However, coupling policy options with politics (decision-coupling) is frequently the most precarious phase of the policy process. In particular, domestic-level power asymmetries that favor structurally entrenched policy opponents will often undermine decision-coupling, even when MLG structures facilitate agenda-coupling.

As such, our findings provide important qualifications to the functionalist expectation that MLG dynamics will facilitate multilevel reinforcement over time "despite temporary setbacks" (Rietig 2020: 56). It also invites further inquiry into the extent to which such claims travel across settings. We attribute the relative absence of multilevel reinforcement in ASEAN to differences between the institutional setup of these two regional organizations – notably the EU's unique approach to "sovereignty pooling" – but also to deeper structural asymmetries that characterize the politics stream in ASEAN, where powerful domestic politico-business coalitions are able to keep both agenda and decision windows firmly shut if policy options are incompatible with their preferences. In so doing, we hope to inspire further research which combines a focus on the micro-aspects of what constitutes "best practice" policy entrepreneurship with an appreciation of deeper systemic factors that work to enable or constrain policy entrepreneurship.

Our study also has important policy implications. Notably, EU policy entrepreneurs have also had to contend with political conflict within and between countries. As our post-Copenhagen case study demonstrates, entrenched status quo interests can also make use of the techniques of policy entrepreneurship to block, delay, or weaken ambitious policy proposals, especially when they are aligned with key insider veto players within domestic political systems. That said, based on our findings, EU-savvy policy entrepreneurs would be well advised to bide their time, in the knowledge that the political landscape might shift, as pressures from above and/or below build up. More broadly, the EU's policy processes have allowed for a much more dynamic competition of ideas; protecting and enhancing existing access points for motivated (if variably resourced) policy entrepreneurs should remain a policy priority.

In contrast, the ASEAN experience demonstrates how a lack of member-state consensus over regional climate policy has effectively excluded ambitious policy entrepreneurs from decision-coupling arenas, evidenced in low and highly selective adoption of open-ended regional policy instruments. As we have evidenced, while NGOs and think tanks in the ASEAN region are certainly significant promoters of the Paris Agreement's goals, they have also been repeatedly marginalized in favor of problem frames and policy ideas compatible with elite preferences. As such, while the MSF offers a valuable conceptual device to capture the contingency of policymaking, we challenge the original MSF's assumption that all three streams are relatively independent of each other and in constant flow. The ASEAN experience in particular suggests that entrenched power asymmetries in the politics stream – stemming from the structure of domestic political economies – pose a formidable policy challenge, placing considerable constraints on the types of frames and ideas which are likely to emerge in the first place. As we have shown, this is also a salient concern in the EU.

Where does that leave us? Global temperatures continue to rise. That said, the real success of the Paris Agreement lies in the construction of an ideational environment which can be leveraged by ambitious policy entrepreneurs on other governance levels. Our study demonstrates that regional organizations have, on occasion, realized this promise, with EU entrepreneurs strongly invested in driving policy ambition at the UNFCCC level and willing to go it alone if necessary. However, carbon-intensive models of national economic development in the ASEAN region (and, we might add, offshoring of the most carbon-intensive and environmentally destructive stages of commodity production to the Global South) have made such an outcome fiendishly difficult elsewhere.

Nevertheless, as pressure mounts, regional organizations are likely to be asked to play an even greater role in mediating climate policy delivery within and across domestic governance systems. Further research is needed on how regional organizations can facilitate policy entrepreneurship, especially in light of new ideological cleavages (Hooghe and Marks 2018), as well as the sheer technical complexity of the decarbonization challenge (Bernstein and Hoffmann 2019). As we approach 2025, buffeted by powerful security and economic crosswinds, a deepening of regional coordination both upward, with the UNFCCC process itself, and downward with increasingly climate-impacted national governments, may yet prove vital to realizing the transformative promise of the Paris Agreement.

Appendix

Country-aggregated data for the EU (plus United Kingdom)

	GDP in 2021 (value in million US$) – World Bank (n.d.)	Population in 2021 (value in thousands) – World Bank (n.d.)	GDP per capita in 2021 (in US$) – World Bank (n.d.)	GHG emissions in 2018 (Mt CO2eq) – UNEP (2021)	GHG emissions in 2018 (as percent of global total emissions) – UNEP (2021)	GHG emissions in 2018 (per capita in metric tons) – UNEP (2021)	Percentage change (2018–1990) – own calculations based on UNEP (2021)
Austria	477,082.47	8,956.28	53,267.9	85.39	0.2	9.76	+4.77
Belgium	599,879.03	11,587.88	51,767.8	130.56	0.3	11.35	-9.33
Bulgaria	80,271.12	6,899.13	11,635.0	59.65	0.1	8.48	-42.09
Croatia	67,837.79	3,899.00	17,398.8	25.71	<0.1%	6.17	-25.26
Cyprus	27,719.34	1,215.59	30,798.5	8.60	<0.1%	7.23	+62.57
Czech Republic	282,340.85	10,703.45	26,378.5	131.41	0.3	12.37	-32.26
Denmark	397,104.34	5,856.73	67,803.0	49.17	0.1	8.54	-30.16
Estonia	36,262.92	1,329.25	27,280.7	25.12	<0.1%	19.22	-38.73
Finland	299,155.24	5,541.70	53,982.6	74.80	0.1	13.5	-9.88
France	2,937,472.76	67,499.34	43,518.5	450.39	0.9	6.9	-18.11
Germany	4,223,116.21	83,129.29	50,801.8	873.60	1.8	10.62	-28.98

Greece	216,240.59	10,664.57	20,276.5	91.86	0.2	8.24	-9.05
Hungary	182,280.52	9,709.89	18,772.7	67.16	0.1	6.93	-30.19
Ireland	498,559.58	5,028.23	99,152.1	67.62	0.1	14.08	+17.19
Italy	2,099,880.20	59,066.22	35,551.3	417.56	0.8	7.04	-21.36
Latvia	38,872.55	1,883.16	20,642.2	12.55	<0.1%	6.51	-55.02
Lithuania	65,503.85	2,795.32	23,433.4	23.45	<0.1%	8.15	-50.21
Luxembourg	86,710.80	639.07	135,682.8	10.46	<0.1%	17.73	-17.64
Malta	17,189.73	516.87	33,257.4	2.06	<0.1%	4.76	-17.93
Netherlands	1,018,007.06	17,533.40	58,061.0	221.90	0.5	12.99	-12.29
Poland	674,048.27	37,781.02	17,840.9	424.96	0.9	11.15	-16.67
Portugal	249,886.46	10,299.42	24,262.2	69.55	0.1	6.76	+19.5
Romania	284,087.56	19,115.15	14,861.9	118.95	0.2	6.08	-51.25
Slovak Republic	114,870.71	5,447.25	21,087.8	47.58	0.1	8.73	-35.78
Slovenia	61,526.33	2,107.01	29,200.8	20.51	<0.1%	9.85	-10.44
Spain	1,425,276.59	47,326.69	30,115.7	349.77	0.7	7.54	+17.37
Sweden	627,437.90	10,415.81	60,239.0	64.59	0.1	6.47	-18.55
United Kingdom	3,186,859.74	67,326.57	47,334.4	463.74	0.9	6.97	-39.85
	17,088,620.74 (EU-27)	446,946.71 (EU-27)	38,234.1 (EU-27)	3,924.93 (EU-27)	~10% (EU-28)	9.52 (EU-27)	-23.32 (EU-28)
				4,388.67 (EU-28)		9.43 (EU-28)	

Country-aggregated data for ASEAN:

	GDP in 2021 (Value in million US$) – World Bank (n.d.)	Population in 2021 (value in thousands) – World Bank (n.d.)	GDP per capita in 2021 (in US$) – World Bank (n.d.)	GHG emissions in 2018 (total in metric tons of GHG) – UNEP (2021)	GHG emissions in 2018 (as percent of global total emissions) – UNEP (2021)	GHG emissions in 2018 per capita in metric tons) – UNEP (2021)	Percentage change (2018–1990) – own calculations based on UNEP (2021)
Brunei Darussalam	14,006.57	441.53	31,722.7	12.80	<0.1%	29.49	+72.27
Cambodia	26,961.06	16,946.45	1,591.0	41.50	<0.1%	2.55	+126.78
Indonesia	1,186,092.99	276,361.79	4,291.8	1,074.19	2.2	4.03	+162
Lao PDR	18,827.15	7,379.36	2,551.3	36.94	<0.1%	5.31	+382.88
Malaysia	372,701.36	32,776.19	11,371.1	324.31	0.7	10.12	+252.51
Myanmar	65,067.81	54,806.01	1,187.2	145.24	0.3	2.7	+118.08
Philippines	394,086.42	111,046.91	3,548.8	237.42	0.5	2.23	+128.29
Singapore	396,986.90	5,453.57	72,794.0	70.52	0.1	12.18	+107.41
Thailand	505,981.66	69,950.84	7,233.4	434.78	0.9	6.28	+109.03
Vietnam	362,637.52	98,168.83	3,694.0	418.80	0.8	4.34	+306.6
	3,343,349.44 (ASEAN total)	673,331,480 (ASEAN total)	4,965.4 (ASEAN average)	2,796.5 (ASEAN total)	~6% (ASEAN total)	7.9 (ASEAN average)	+166.09 (ASEAN total)

Bibliography

Abnett, K. and Twidale, S. (2021). EU Proposes World's First Carbon Border Tax for Some Imports. *Reuters*, July 14. www.reRuters.com/business/sustainable-business/eu-proposes-worlds-first-carbon-border-tax-some-imports-2021-07-14/.

Abrams, S., Dietzel, A., Hammond, M. et al. (2020). Just Transition: Pathways to Socially Inclusive Decarbonisation. *COP26 Universities Network Briefing*. October 2020. www.ucl.ac.uk/global-governance/sites/global-governance/files/cop26_just_transition_policy_paper.pdf.

Acharya, A. (2003). Regional Institutions and Asian Security Order; Norms, Power, and Prospects for Peaceful Change. In M. Alagappa (ed.), *Asian Security Order: Instrumental and Normative Features*. Stanford, CA: Stanford University Press, 210–240.

Ackrill, R. and Kay, A. (2011). Multiple Streams in EU Policy-Making: The Case of the 2005 Sugar Reform. *Journal of European Public Policy*, 18(1), 72–89. https://doi.org/10.1080/13501763.2011.520879.

Ackrill, R., Kay, A. and Zahariadis, N. (2013). Ambiguity, Multiple Streams, and EU Policy. *Journal of European Public Policy*, 20 (6), 871–887. https://doi.org/10.1080/13501763.2013.781824.

Activism in Crisis (2020). About Activism in Crisis. www.activismincrisis.com/about-activism-in-crisis.

Asian Development Bank (ADB) (2017). A Region at Risk: The Human Dimensions of Climate Change in Asia and the Pacific. www.adb.org/publications/region-at-risk-climate-change.

Asian Development Bank (ADB) (2021). Climate Change Financing at ADB. https://data.adb.org/dashboard/climate-change-financing-adb.

ASEAN Energy Market Integration (AEMI) (2015). Energy Security and Connectivity: The Nordic and European Union Approaches. https://esi.nus.edu.sg/docs/default-source/doc/proceedings-of-aemi-forum-2015.pdf.

Agustinus, M. (2016). Sudah 5 Kali Menteri ESDM Tegur Keras PLN. *Detikfinance*, July 22. https://finance.detik.com/energi/d-3259177/sudah-5-kali-menteriesdm-tegur-keras-pln.

Ahmed, T., Mekhilef, T., Shah, R. et al. (2017a). ASEAN Power Grid: A Secure Transmission Infrastructure for Clean and Sustainable Energy for South-East Asia. *Renewable and Sustainable Energy Reviews*, 67, 1420–1435.

Ahmed, T., Mekhilef, T., Shah, R. and Mithulananthan, N. (2017b). Investigation into Transmission Options for Cross-border Power Trading in ASEAN Power

Grid. *Energy Policy*, 108, 91–101. https://doi.org/10.1016/j.enpol.2017.05
.020.

Aklin, M. and Mildenberger, M. (2020). Prisoners of the Wrong Dilemma: Why
Distributive Conflict, Not Collective Action, Characterizes the Politics of
Climate Change. *Global Environmental Politics*, 20 (4), 4–27. https://doi
.org/10.1162/glep_a_00578.

Al-Fadhat, F. (2020). The Internationalisation of Capital and the
Transformation of Statehood in Southeast Asia. In T. Carroll, S. Hameiri
and L. Jones (eds.), *The Political Economy of Southeast Asia*. 4th ed. Cham:
Springer International, 177–198.

Andrews-Speed, P. (2016). *Connecting ASEAN through the Power Grid: Next
Steps*. NUS Energy Studies Institute, Policy Brief No. 11. https://esi.nus.edu
.sg/docs/default-source/esi-policy-briefs/connecting-asean-through-the-
power-grid-next-steps.pdf?sfvrsn=4.

Arinaldo, D. and Adiatma, J. C. (2019). Dinamika Batubara Indonesia: Menuju
Transisi Energi yang Adil. Jakarta: Institute for Essential Services Reform
(IESR). http://iesr.or.id/wp-content/uploads/2019/04/SPM-bahasa-lowres
.pdf.

ASEAN (2018). About ASEAN Cooperation on Environment. https://environ
ment.asean.org/about-asean-cooperation-on-environment/.

The ASEAN Post (2019). ASEAN's Renewable Energy Challenges.
December 9. https://theaseanpost.com/article/aseans-renewableenergy-
challenges.

Badan Restorasi Gambut (2016). Rencana Strategis Badan Restorasi Gambut
2016–2020. https://brg.go.id/files/RENSTRA%20BRG%202016-2020%20
(November%202016).pdf.

Barbière, C. (2019). Paris, Berlin Divided Over Nuclear's Recognition as Green
Energy. *EURACTIV*, November 27 . www.euractiv.com/section/energy-envir
onment/news/france-and-germany-divided-over-nuclears-inclusion-in-eus-
green-investment-label/.

Barnes, P. M. (2010). The Role of the Commission of the European Union:
Creating External Coherence from Internal Diversity. In R. K. W. Wurzel and
J. Connelly (eds.), *The European Union as a Leader in International Climate
Change Politics*. London: Routledge, 41–57.

Baumgartner, F. and Jones, B. (2009). *Agendas and Instability in American
Politics*. 2nd ed. Chicago: University of Chicago Press.

Becker, M. and Traufetter, G. (2016). How Officials Ignored Years of Emissions
Evidence. *Spiegel International*, August 19. www.spiegel.de/international/
business/volkswagen-how-officials-ignored-years-of-emissions-evidence-
a-1108325.html.

Becker, P. (2019). The Reform of European Cohesion Policy or How to Couple the Streams Successfully. *Journal of European Integration*, 41(2), 147–168. https://doi.org/10.1080/07036337.2018.1553964.

Berliner, D. and Prakash, A. (2014). Public Authority and Private Rules: How Domestic Regulatory Institutions Shape the Adoption of Global Private Regimes. *International Studies Quarterly*, 58(4), 793–803. https://doi.org/10.1111/isqu.12166.

Bernstein, S. and Hoffmann, M. (2019). Climate Politics, Metaphors and the Fractal Carbon Trap. *Nature Climate Change*, 9, 919–925.

Betsill, M. M. (2007). Regional Governance of Global Climate Change: The North American Commission for Environmental Cooperation. *Global Environmental Politics*, 7(2), 11–27. https://doi.org/10.1162/glep.2007.7.2.11.

Bocquillon, P. and Dobbels, M. (2013). An Elephant on the 13th Floor of the Berlaymont? European Council and Commission Relations in Legislative Agenda Setting. *Journal of European Public Policy*, 21(1), 20–38. https://doi.org/10.1080/13501763.2013.834548.

Börzel, T. A. and van Hüllen, V. (2015). Towards a Global Script? Governance Transfer by Regional Organizations. In T. A. Börzel and V. van Hüllen (eds.), *Governance Transfer by Regional Organizations*. Houndmills: Palgrave Macmillan, 3–21.

Bosnia, T. (2018). Sandiaga Uno Masih Punya Saham Bernilai Rp 2, 59 T di Saratoga. *CNBC Indonesia*, October 5. www.cnbcindonesia.com/market/20181005142744-17-36216/sandiaga-unomasih-punya-saham-bernilai-rp-259-t-di-saratoga.

Brauers, H. and Oei, P.-Y. (2020). The Political Economy of Coal in Poland: Drivers and Barriers for a Shift Away from Fossil Fuels. *Energy Policy*, 144 (111621), 1–12. https://doi.org/10.1016/j.enpol.2020.111621.

Braun, D. (2021). The Europeanness of the 2019 EP Elections and the Mobilizing Power of European Issues. *Politics*, 41(4), 1–16. https://doi.org/10.1177/0263395721992930.

Braun, M. (2009). The Evolution of Emissions Trading in the European Union – The Role of Policy Networks, Knowledge and Policy Entrepreneurs. *Accounting, Organizations and Society*, 34(3–4), 469–487. https://doi.org/10.1016/j.aos.2008.06.002.

Büntgen, U., Urban, O., Krusic, P. J. et al. (2021). Recent European Drought Extremes beyond Common Era Background Variability. *Nature Geoscience*, 14, 190–196. https://doi.org/10.1038/s41561-021-00698-0.

Bürgin, A. (2014). National Binding Renewable Energy Targets for 2020, But Not for 2030 Anymore: Why the European Commission Developed from a Supporter to a Brakeman. *Journal of European Public Policy*, 22(5), 690–707.

Burns, C. (2019). In the Eye of the Storm? The European Parliament, the Environment and the EU's Crises. *Journal of European Integration*, 41(3), 311–327, https://doi.org/10.1080/07036337.2019.1599375.

Cairney, P. (2018). Three Habits of Successful Policy Entrepreneurs. *Policy & Politics*, 46(2), 199–215. https://doi.org/10.1332/030557318X152300567 71696.

Cairney, P. and Jones, M. D. (2016). Kingdon's Multiple Streams Approach: What is the Empirical Impact of this Universal Theory? *Policy Studies Journal*, 44(1), 37–58. https://doi.org/10.1111/psj.12111.

Cairney, P. and Zahariadis, N. (2016). Multiple Streams Approach: A Flexible Metaphor Presents an Opportunity to Operationalize Agenda Setting Processes. In N. Zahariadis (ed.), *Handbook of Public Policy Agenda Setting*. Cheltenham: Edward Elgar, 87–105. https://doi.org/10.4337/978178 4715922.00014.

Cañete, M. A. (2015). Historic Climate Deal in Paris: Speech by Commissioner Miguel Arias Cañete at the Press Conference on the Results of COP21 Climate Conference in Paris. *European Commission*. https://ec.europa.eu/ commission/presscorner/detail/en/SPEECH_15_6320.

Carey, E. (2015). The European Union's 2030 Climate Energy Package and the COP 21. Translated version of "Le Paquet Energie-Climat 2030 de l'Union Européenne et la Conférence de Paris sur le Climat (COP 21)." *Annuaire Français des Relations Internationales*, 16. www.afri-ct.org/wpcontent/ uploads/2015/11/2848_-_Article_Thucydide_EU_EnergyClimate_2030_ Package_and_the_COP_21.pdf.

Carrington, D. (2013). Angela Merkel "Blocks" EU Plan on Limiting Emissions from New Cars. *The Guardian*, June 28. www.theguardian.com/environ ment/2013/jun/28/angela-merkel-eu-caremissions.

Carroll, T. (2020). The Political Economy of Southeast Asia's Development from Independence to Hyperglobalisation. In T. Carroll, S. Hameiri and L. Jones (eds.), *The Political Economy of Southeast Asia*. 4th ed. Cham: Springer International, 35–84.

Carter, N. and Childs, M. (2018). Friends of the Earth as a Policy Entrepreneur: "The Big Ask" Campaign for a UK Climate Change Act. *Environmental Politics*, 27(6), 994–1013. https://doi.org/10.1080/09644016.2017.1368151.

Carty, T., Kowalzig, J. and Zagema, B. (2020). Climate Finance Shadow Report 2020: Assessing Progress towards the $100 billion Commitment. *Oxfam International*. https://oxfamilibrary.openrepository.com/handle/10546/ 621066.

Ciplet, D. and Roberts, J. T. (2017). Climate Change and the Transition to Neoliberal Environmental Governance. *Global Environmental Change*, 46, 148–156. https://doi.org/10.1016/j.gloenvcha.2017.09.003.

Clark, R., Zucker, N. and Urpelainen, J. (2020). The Future of Coal-Fired Power Generation in Southeast Asia. *Renewable and Sustainable Energy Reviews*, 121(April), 109650.

Coen, D., Katsaitis, A. and Vannoni, M. (2021). *Business Lobbying in the European Union*. Oxford: Oxford University Press.

Cohen, M. D., March, J. G. and Olsen, J. P. (1972). A Garbage Can Model of Organizational Choice. *Administrative Science Quarterly*, 17(1), 1–25. https://doi.org/10.2307/2392088.

Copeland, P. and James, S. (2013). Policy Windows, Ambiguity and Commission Entrepreneurship: Explaining the Relaunch of the European Union's Economic Reform Agenda. *Journal of European Public Policy*, 21(1), 1–19. https://doi.org/10.1080/13501763.2013.800789.

Curry, D. (2015). *Network Approaches to Multi-Level Governance: Structures, Relations and Understanding Power between Levels*. Basingstoke: Palgrave Macmillan.

Dagnet, Y., Cogswell, N., Grinspan, D., Reichart, E. and Drew, D. (2019a). *Date and Ambition Loops for Enhanced Climate Action: Potential Drivers and Opportunities in Asia*. Washington, DC: World Resource Institute. www.wri.org/publication/data-and-ambition-loops.

Dagnet, Y., Cogswell, N. and Mendoza, J. M. (2019b). INSIDER: How Can Governments and Businesses Accelerate Climate Action? Through "Data and Ambition Loops." *World Resource Institute Indonesia*, September 13. https://wri-indonesia.org/en/blog/insider-how-can-governments-and-businesses-accelerate-climate-action-through-%E2%80%9Cdata-and-ambition.

Dauvergne, P. (2018). The Global Politics of the Business of "Sustainable" Palm Oil. *Global Environmental Politics*, 18(2), 34–52. https://doi.org/10.1162/glep_a_00455.

Delbeke, J. and Vis, P., eds. (2015). *EU Climate Policy Explained*. Abingdon: Routledge.

Domorenok, E. (2019). Voluntary Instruments for Ambitious Objectives? The Experience of the EU Covenant of Mayors. *Environmental Politics*, 28(2), 293–314. https://doi.org/10.1080/09644016.2019.1549777.

Dreger, J. (2014). *The European Commission's Energy and Climate Policy: A Climate for Expertise?* Basingstoke: Palgrave Macmillan.

Duijndam, S. and van Beukering, P. (2020). Understanding Public Concern about Climate Change in Europe, 2008–2017: The Influence of Economic

Factors and Right-wing Populism. *Climate Policy*, 21(3), 353–367. https://doi.org/10.1080/14693062.2020.1831431.

Dunai, M. (2019). Hungary Lists Conditions, "Won't Sign Blank Cheque" on EU Climate Deal. *Reuters*, December 12. www.reuters.com/article/us-climate-change-eu-hungary-idUSKBN1YG0XU.

Dupont, C. and Oberthür, S. (2015). The European Union. In K. Bäckstrand and E. Lövbrand (eds.), *Research Handbook on Climate Governance*. Cheltenham: Edward Elgar, 224–236.

Dupont, C., Oberthür, S. and von Homeyer, I. (2020). The Covid-19 Crisis: A Critical Juncture for EU Climate Policy Development? *Journal of European Integration*, 42(8), 1095–1110. https://doi.org/10.1080/07036337.2020.1853117.

Duwe, M., and Evans, N. (2020). *Climate Laws in Europe: Good Practices in Net-Zero Management.* Berlin: Ecologic. February. www.ecologic.eu/17233.

Eckersley, R. (2020). Greening States and Societies: From Transitions to Great Transformations, *Environmental Politics*, 1–21. https://doi.org/10.1080/09644016.2020.1810890.

Economic Research Institute for ASEAN and East Asia (ERIA) (2015). Study on Effective Power Infrastructure Investment through Power Grid Interconnections in East Asia. www.eria.org/publications/study-on-effective-power-infrastructure-investment-through-power-grid-interconnections-in-east-asia/.

Economic Research Institute for ASEAN and East Asia (ERIA) (2018). Distributed Energy Systems in South East Asia. www.eria.org/uploads/media/Distributed_Energy_System_in_Southeast_Asia_book.pdf.

Elkind, J. and Bednarz, D. (2020). *Warsaw, Brussels, and Europe's Green Deal: Challenges and Opportunities in 2020.* Center on Global Energy Policy, Columbia School of International and Public Affairs. www.energypolicy.columbia.edu/sites/default/files/file-uploads/EUGreenDeal_CGEP_Commentary_072820-4.pdf.

Environmental Insights (2020). Carbon Pricing History and Future: An Interview with Jos Delbeke. Podcast Transcript. *Harvard Environmental Economics Program.* https://heep.hks.harvard.edu/files/heep/files/jos_delbeke_environmental_insights_podcast_transcript_01.pdf.

Eurobarometer (2019). Special Eurobarometer 490: Climate Change. European Commission, April . https://ec.europa.eu/clima/sites/clima/files/support/docs/report_2019_en.pdf.

Eurobarometer (2021). *Special Eurobarometer 513: Climate Change.* Full Report. Brussels: European Commission. https://europa.eu/eurobarometer/surveys/detail/2273.

European Commission (2010). Analysis of Options to Move beyond 20% Greenhouse Gas Emission Reductions and Assessing the Risk of Carbon Leakage. COM(2010) 265 final, May 26. https://eur-lex.europa.eu/legal-con tent/EN/TXT/PDF/?uri=CELEX:52010DC0265&from=HR.

European Commission (2018). A Clean Planet for All: A European Strategic Long-term Vision for a Prosperous, Modern, Competitive and Climate Neutral Economy. COM(2018) 773 final. *European Commission*, November 28. https://eur-lex.europa.eu/legal-content/EN/TXT/PDF/? uri=CELEX:52018DC0773&from=EN.

European Environment Agency (EEA) (1999). Environment in the European Union at the Turn of the Century. www.eea.europa.eu/publications/92-9157-202-0.

European Environment Agency (EEA) (2013). *Trends and Projections in Europe 2013. Tracking Progress towards Europe's Climate and Energy Targets until 2020.* Executive Summary. EEA Report No. 10/2013. www .eea.europa.eu/publications/trends-and-projections-2013.

European Parliament (2020). EU Climate Law: MEPs Want to Increase 2030 Emissions Reduction Target to 60%. Press Release, October 8. www.euro parl.europa.eu/news/en/press-room/20201002IPR88431/eu-climate-law-meps-want-to-increase-2030-emissions-reduction-target-to-60.

Fabbrini, F. (2015). States' Equality v States' Power: The Euro-crisis, Inter-state Relations and the Paradox of Domination. *Cambridge Yearbook of European Legal Studies*, 17, 3–35. https://doi.org/10.1017/cel.2014.1.

Farand, C. (2019). Climate a "Signature Issue" as Ursula von der Leyen Anointed EU Chief. *Climate Home News*, July 16. www.climatechange news.com/2019/07/16/climate-plays-decisive-role-ursulavon-der-leyen-annointed-eu-chief/.

Fernandez, R. M. (2018). Conflicting Energy Policy Priorities in EU Energy Governance. *Journal of Environmental Studies and Sciences*, 8, 239–248. https://doi.org/10.1007/s13412-018-0499-0.

Fischer, S. and Geden, O. (2015). The Changing Role of International Negotiations in EU Climate Policy. *The International Spectator*, 50(1), 1–7.

Fitch-Roy, O., Bensons, D. and Mitchell, C. (2018). Wipeout? Entrepreneurship, Policy Interaction and the EU's 2030 Renewable Energy Target. *Journal of European Integration*, 41(1), 87–103. https://doi.org/10.1080/07036337.2018.1487961.

Forest Watch Indonesia, Mining Advocacy Network (JATAM), POKA30, Trend Asia and The Indonesian Forum for Environment (WALHI) (2019). Ibu Kota Baru Buat Siapa? Action Research Report. https://fwi.or.id/publikasi/ibu-kota-baru-buat-siapa/.

Free Malaysia Today (2022). Lift Ban on Renewable Energy Exports, Yeo Urges Govt. *Free Malaysia Today*, May 24. www.freemalaysiatoday.com/category/nation/2022/05/24/lift-ban-on-renewable-energy-exports-yeo-urges-govt/.

Friends of the Earth (2022). No Consent: Astra Agro Lestari's Land Grab in Central and West Sulawesi, Indonesia. https://foe.org/resources/astra-agro-lestari//.

Fuchs, D. and Feldhoff, B. (2016). Passing the Scepter, not the Buck. *Journal of Sustainable Development*, 9(6), 58–74.

Gaskell, J. C. (2015). The Role of Markets, Technology, and Policy in Generating Palm-Oil Demand in Indonesia. *Bulletin of Indonesian Economic Studies*, 51(1), 29–45.

Gellert, P. K. (2015). Palm Oil Expansion in Indonesia: Land Grabbing as Accumulation by Dispossession. In J. Shefner (ed.), *States and Citizens: Accommodation, Facilitation and Resistance to Globalization: v.34 (Current Perspectives in Social Theory 34)*. Bingley: Emerald Group, 65–99.

Gellert, P. K. (2020). The Political Economy of Environmental Degradation and Climate Disaster in Southeast Asia. In T. Carroll, S. Hameiri and L. Jones (eds.), *The Political Economy of Southeast Asia*. 4th ed. Cham: Springer International, 367–387.

Gerard, K. (2014). *ASEAN's Engagement of Civil Society: Regulating Dissent*. 1st ed. London: Palgrave MacMillan.

Goron, C. (2014). EU–ASEAN Relations in the Post-2015 Climate Regime: Exploring Pathways for Top-down and Bottom-up Climate Governance. In P. Rueppel and W. Hofmeister (eds.), *Climate Change Diplomacy: The Way Forward for Asia and Europe*. Singapore: Konrad Adenauer Stiftung, 101–130.

Green, J. (2020). Less Talk, More Walk: Why Climate Change Demands Activism in the Academy. *Daedalus*, 149(4), 151–162. https://doi.org/10.1162/daed_a_01824.

Greenpeace (2017). Six Years of Moratorium: How Much of Indonesia's Forests Have Been Protected? *Greenpeace Southeast Asia*. www.greenpeace.org/southeastasia/press/684/six-years-of-moratorium-howmuch-of-indonesias-forests-protected/.

Greenpeace (2019). Indonesian Government Actively Blocking Efforts to Reform Palm Oil Industry. *Greenpeace Southeast Asia*. Press Release, May 16. www.greenpeace.org/southeastasia/press/2448/indonesian-government-actively-blocking-efforts-to-reform-palm-oil-industry/.

Greenpeace (2020). Southeast Asia Power Sector Scorecard. www.greenpeace.org/static/planet4-southeastasia-stateless/2020/09/8f7592a7-gpsea-southeast-asia-power-sector-scorecard-report-170920-fv7.pdf.

Greenpeace (2021). Investment Trends: Making the Case for SE Asian Renewables. January 5. www.greenpeace.org/eastasia/blog/6278/invest ment-trends-making-the-case-for-se-asian-renewables/.

Grubb, M., Laing, T., Sato, M. and Comberti, C. (2012). *Analyses of the Effectiveness of Trading in EU-ETS.* Working Paper. London: Climate Strategies. https://climatestrategies.org/wp-content/uploads/2014/11/cs-effectiveness-of-ets.pdf.

Guild, J. (2019). Feed-in-Tariffs and the Politics of Renewable Energy in Indonesia and the Philippines. *Asia & the Pacific Policy Studies,* 6(3), 417–431.

Guild, J. (2020). The Political and Institutional Constraints on Green Finance in Indonesia. *Journal of Sustainable Finance & Investment,* 10(2), 157–170.

Hameiri, S. and Jones, L. (2013). The Politics and Governance of Non-Traditional Security. *International Studies Quarterly,* 57(3), 462–473.

Hameiri, S. and Jones, L. (2015). *Governing Borderless Threats: Non-Traditional Security and the Politics of State Transformation.* Cambridge: Cambridge University Press.

Hameiri, S. and Jones, L. (2017). Beyond Hybridity to the Politics of Scale: International Intervention and "Local" Politics. *Development and Change,* 48 (1), 54–77. https://doi.org/10.1111/dech.12287.

Han, K. (2020). Climate Change Activists Test Strict Singapore Protest Laws. *Aljazeera,* April 10. www.aljazeera.com/news/2020/04/climate-changeacti vists-test-strict-singapore-protest-laws-200409082537461.html.

Harcourt, A. (2016). Communications Policy in the European Union: The UK as a Policy Entrepreneur. In N. Zahariadis (ed.), *Handbook of Public Policy Agenda Setting.* Cheltenham: Edward Elgar, 332–347.

Harrison, N. and Geyer, R. R. (2019). The Complexity of the Governance of Climate Change. *International Studies Review,* 22(4), 1028–1031. https://doi.org/10.1093/isr/viz005.

Hatcher, P. (2020). The Political Economy of Southeast Asia's Extractive Industries: Governance, Power Struggles and Development Outcomes. In T. Carroll, S. Hameiri and L. Jones (eds.), *The Political Economy of Southeast Asia.* 4th ed. Cham: Springer International, 317–339.

He, A. J. and Ma, L. (2019). Corporate Policy Entrepreneurship and Cross-boundary Strategies: How a Private Corporation Champions Mobile Healthcare Payment Innovation in China? *Public Administration and Development,* 40(1), 76–86. https://doi.org/10.1002/pad.1868.

Herweg, N., Huß, C. and Zohlnhöfer, R. (2015). Straightening the Three Streams: Theorising Extensions of the Multiple Streams Framework. *European Journal of Political Research,* 54(3), 435–449. https://doi.org/10.1111/1475-6765.12089.

Herweg, N., Zahariadis, N. and Zohlnhöfer, R. (2017). The Multiple Streams Framework: Foundations, Refinements and Empirical Applications. In C. M. Weible and P. A. Sabatier (eds.), *Theories of the Policy Process*. Boulder, CO: Westview Press, 17–54.

Hicks, C. (2019). Viet Nam Becomes the First Country in Asia-Pacific to Fulfill the Warsaw Framework for REDD+. *UN-REDD Programme*. February 1. www.un-redd.org/post/2019/02/25/viet-nam-submits-its-first-summary-ofin formation-on-safeguards-to-the-unfccc-and-release.

Hicks, R. (2020). Singapore's Covid-19 Economic Recovery Team Criticised for Excluding Sustainable Business Voices, Women. *Eco-Business*, June 11. www.eco-business.com/news/singapores-covid-19-economic-recovery-teamcriticised-for-excluding-sustainable-business-voices-women/.

Hix, S. (2007). The European Union as a Polity (I). In K. Joergensen, M. Pollack and B. Rosamond (eds.), *The Sage Handbook of European Union Politics*. London: Sage, 141–158.

Hix, S. and Marsh, M. (2007). Punishment or Protest? Understanding European Parliament Elections. *The Journal of Politics*, 69(2), 495–510.

Hooghe, L. and Marks, G. (2003). Unraveling the Central State, but How? Types of Multi-level Governance. *American Political Science Review*, 97, 233–243. https://doi.org/10.1017/S0003055403000649.

Hooghe, L. and Marks, G. (2008). A Postfunctionalist Theory of European Integration: From Permissive Consensus to Constraining Dissensus. *British Journal of Political Science*, 39(1), 1–23. https://doi.org/10.1017/S0007123 408000409.

Hooghe, L. and Marks, G. (2018). Cleavage Theory Meets Europe's Crises: Lipset, Rokkan, and the Transnational Cleavage. *Journal of European Public Policy*, 25(1), 109–135. https://doi.org/10.1080/13501763.2017.1310279.

Howlett, M., McConnell, A. and Perl, A. (2015). Streams and Stages: Reconciling Kingdon and Policy Process Theory. *European Journal of Political Research*, 54 (3), 419–434. https://doi.org/10.1111/1475-6765.12064.

Human Rights Watch (2020). Cambodia: Thai Activist Abducted in Phnom Penh. June 5. www.hrw.org/news/2020/06/05/cambodia-thai-activ ist-abductedphnom-penh.

Huynh, N. T., Lin, W., Ness, L. R., Occeña-Gutierrez, D., and Trần, X. D. (2014). Climate Change and Its Impact on Cultural Shifts in East and Southeast Asia. In Jim Norwine (ed), *A World After Climate Change and Culture-Shift*. Dordrecht: Springer, 245–302.

IEA (2020). "Thailand". https://www.iea.org/countries/thailand.

International Energy Agency (IEA) (2021). Global Electricity Demand Is Growing Faster than Renewables, Driving Strong Increase in Generation from Fossil Fuels. July 15. www.iea.org/news/global-electricity-demand-is-growing-faster-than-renewables-driving-strong-increase-in-generation-from-fossil-fuels.

Institute for European Environmental Policy (IEEP) (2020). More than Half of All CO2 Emissions since 1751 Emitted in the Last 30 years. April 29. https://ieep.eu/news/more-than-half-of-all-co2-emissions-since-1751-emitted-in-the-last-30-years.

Intergovernmental Panel on Climate Change (IPCC) (2001). *Third Assessment Report [Synthesis].* . www.ipcc.ch/site/assets/uploads/2018/05/SYR_TAR_full_report.pdf.

Intergovernmental Panel on Climate Change (IPCC) (2018). Global Warming of 1.5 °C [Special Report]. . www.ipcc.ch/sr15/.

Intergovernmental Panel on Climate Change (IPCC) (2022). Climate Change 2022: Mitigation of Climate Change. www.ipcc.ch/report/ar6/wg3/.

The Jakarta Post (2014). Salim, Wilmar to Acquire Aussie Bread Firm. *The Jakarta Post*, May 28, 2014. www.thejakartapost.com/news/2014/05/28/salimwilmar-acquire-aussie-bread-firm.html.

The Jakarta Post (2017). Renewable Energy Regulation Repeats Old Mistakes: Association. *The Jakarta Post*, February 6, 2017. https://www.thejakartapost.com/news/2017/02/06/renewable-energy-regulation-repeats-old-mistakes-association.html

Jagers, S. C. and Stripple, J. (2003). Climate Governance beyond the State. *Global Governance*, 9(3), 385–399.

Jänicke, M. (2017). The Multi-level System of Global Climate Governance – the Model and its Current State. *Environmental Policy and Governance*, 27(2), 108–121. https://doi.org/10.1002/eet.1747.

Jänicke, M. and Quitzow, R. (2017). Multi-level Reinforcement in European Climate and Energy Governance: Mobilizing Economic Interests at the Sub-national Levels. *Environmental Policy and Governance*, 27(2), 122-136. https://doi.org/10.1002/eet.1748.

Jänicke, M. and Wurzel, R. K. W. (2018). Leadership and Lesson-drawing in the European Union's Multilevel Climate Governance System. *Environmental Politics*, 28(1), 22–42. https://doi.org/10.1080/09644016.2019.1522019.

JATAM (Mining Advocacy Network) (2019). Daftar Aktor Yang Mendapat Untung. www.jatam.org/wpcontent/uploads/2019/12/PENGUSAHA-TAMBANG-OK.png.

Jeffrey, C. and Peterson, J. (2020). "Breakthrough" Political Science: Multi-level Governance – Reconceptualising Europe's Modernised Polity. *The British Journal of Politics and International Relations*, 22(4), 753–766. https://doi.org/10.1177/1369148120959588.

Jessop, S. (2021). EU Launches Green Bond Framework to Help it Meet Climate Goals. *Reuters*, July 6. www.reuters.com/business/sustainable-business/eu-launches-green-bond-framework-help-it-meet-climate-goals-2021-07-06/.

Jones, L. (2015). Explaining the Failure of the ASEAN Economic Community: The Primacy of Domestic Political Economy. *The Pacific Review*, 29(5), 647–670.

Jong, H. N. (2020). Indonesian lawmakers Push to Pass Deregulation Bills as COVID-19 Grips Country. *Mongabay*, April 6 2020. https://news.mongabay.com/2020/04/indonesia-parliament-dpr-omnibus-bill-mining-covid19/.

Jordan, A. and Huitema, D. (2014). Policy Innovation in a Changing Climate: Sources, Patterns and Effects. *Global Environmental Change*, 29, 387–394. https://doi.org/10.1016/j.gloenvcha.2014.09.005.

Jordan, A. J. and Moore, B. (2020). *Durable by Design? Policy Feedback in a Changing Climate*. Cambridge: Cambridge University Press.

Jordan, A., Huitema, D. and van Asselt, H. (2010). Climate Change Policy in the European Union: An Introduction. In A. Jordan, D. Huitema, H. van Asselt, T. Rayner and F. Berkhout (eds.), *Climate Change Policy in the European Union: Confronting the Dilemmas of Mitigation and Adaptation?* Cambridge: Cambridge University Press, 3–25.

Karim, M. F. (2019). State Transformation and Cross-border Regionalism in Indonesia's Periphery: Contesting the Centre. *Third World Quarterly*, 40(8), 1554–1570. https://doi.org/10.1080/01436597.2019.1620598.

Katzenstein, P. J. and Shiraishi, T. (1997). *Network Power: Japan and Asia*. Ithaca, NY: Cornell University Press.

Kern, K. (2018). Cities as Leaders in EU Multilevel Climate Governance: Embedded Upscaling of Local Experiments in Europe. *Environmental Politics*, 28(1), 125–145. https://doi.org/10.1080/09644016.2019.1521979.

Kim, S. Y. (2011). Public Perceptions of Climate Change and Support for Climate Policies in Asia: Evidence from Recent Polls. *The Journal of Asian Studies*, 70(2), 319–331. https://doi.org/10.1017/S0021911810 00064.

Kingdon, J. W. (1984). *Agendas, Alternatives and Public Policies*. 1st ed. New York: Harper Collins.

Kingdon, J. W. (1995). *Agendas, Alternatives, and Public Policies*. 2nd ed. New York: Harper Collins.

Kingdon, J. W. (2003). *Agendas, Alternatives, and Public Policies*. 2nd ed. New York: Longman.

Kingdon, J. W. (2011). *Agendas, Alternatives and Public Policy*. Boston, MA: Longman.

Koop, C., Reh, C. and Bressanelli, E. (2021). Agenda-setting under Pressure: Does Domestic Politics Influence the European Commission? *European Journal of Political Research*, 1–21. https://doi.org/10.1111/1475-6765 .12438.

Kreppel, A. and Webb, M. (2019). European Parliament Resolutions – Effective Agenda Setting or Whistling into the Wind? *Journal of European Integration*, 41(3), 383–404. https://doi.org/10.1080/07036337.2019 .1599880.

Kurohi, R. (2021). Singapore Parliament Declares Climate Change a Global Emergency. The Straits Time, February 1. www.straitstimes.com/singapore/ politics/singapore-parliament-declares-climate-change-a-global-emergency.

Laffan, B. (1997). From Policy Entrepreneur to Policy Manager: The Challenge Facing the European Commission. *Journal of European Public Policy*, 4(3), 422–438.

Laffan, B. and O'Mahony, J. (2008). "Bringing Politics Back In." Domestic Conflict and the Negotiated Implementation of EU Nature Conservation Legislation in Ireland. *Journal of Environmental Policy and Planning*, 10 (2), 175–197. https://doi.org/10.1080/15239080801928428.

Larik, J. (2019). Regional Organizations' Relations with International Institutions: The EU and ASEAN Compared. In C. A. Bradley (ed.), *The Oxford Handbook of Comparative Foreign Relations Law*. Oxford: Oxford University Press, 447–464.

Laville, S. (2019). Fossil Fuel Big Five "Spent €251 m Lobbying EU" since 2010. *The Guardian*, October 24. www.theguardian.com/business/2019/oct/24/fossil-fuel-big-five-spent-251mlobbying-european-union-2010-climate-crisis.

Lee, M. (2014). *EU Environmental Law, Governance and Decision-Making*. 2nd ed. Oxford: Hart.

Lenschow, A. and Sprungk, C. (2010). The Myth of a Green Europe. *Journal of Common Market Studies*, 48(1), 133–154. https://doi.org/10.1111/j.1468-5965.2009.02045.x.

Marks, G. (1993). Structural Policy and Multilevel Governance in the EC. In A. Cafruny and G. Rosenthal (eds.), *The State of the European Community: The Maastricht Debate and Beyond*. Boulder, CO: Lynne Rienner, 391–411.

Marks, G. (1996). An Actor-Centred Approach to Multi-Level Governance. *Regional and Federal Studies*, 6(2), 20–38. https://doi.org/10.1080/ 13597569608420966.

Martín-Cubas, J., Bodoque, A., Pavía, J. M., Tasa, V. and Veres-Ferrer, E. (2019). The "Big Bang" of the Populist Parties in the European Union: The 2014 European Parliament Election. *Innovation: The European Journal of Social Science Research*, 32(2), 168–190.

Mathiesen, K. and Oroschakoff, K. (2020). "Rather Exhausted" EU Leaders Sign Off on Higher 2030 Climate Goals. *Politico*, December 11. www .politico.eu/article/eu-leaders-agree-to-cut-emissions-55-percent-by-2030/.

Maulia, E. (2021). Indonesia's Switch to Solar is all Charged Up with Nowhere to Go. *Financial Times*, November 7. https://archive.ph/Smogo#selection-1509.0-1509.64.

McCauley, D. and Heffron, R. (2018). Just Transition: Integrating Climate, Energy and Environmental Justice. *Energy Policy*, 119, 1–7. https://doi.org/10.1016/j.enpol.2018.04.014.

Meinshausen, M., Lewis, J., McGlade, C. et al. (2022). Realization of Paris Agreement Pledges May Limit Warming Just below 2°C. *Nature*, 604, 304–316.

Mintrom, M. (2019). So You Want to be a Policy Entrepreneur? *Policy Design and Practice*, 2(4), 307–323. https://doi.org/10.1080/25741292.2019 .1675989.

Mongabay (2020). American Journalist Philip Jacobson Freed after Prolonged Detention in Indonesia. *Mongabay Environmental News*, January 31. https://news.mongabay.com/2020/01/american-journalist-philip-jacobson-freed-afterprolonged-detention-in-indonesia/.

Mukherjee, I. and Giest, S. (2017). Designing Policies in Uncertain Contexts: Entrepreneurial Capacity and the Case of the European Emission Trading Scheme. *Public Policy and Administration*, 34(3), 262–286. https://doi.org/10.1177/0952076717730426.

Mukherjee, I. and Howlett, M. (2015). Who is a Stream? Epistemic Communities, Instrument Constituencies and Advocacy Coalitions in Public Policy-making. *Politics and Governance*, 3(2), 65–75. https://doi .org/10.17645/pag.v3i2.290.

Müller, H. and Van Esch, F. A. W. J. (2019). The Contested Nature of Political Leadership in the European Union: Conceptual and Methodological Cross-Fertilisation. *West European Politics*, 43(5), 1051–1071. https://doi .org/10.1080/01402382.2019.1678951.

Munta, M. (2020). *The European Green Deal. A Game Changer or Simply a Buzzword?* Zagreb: Friedrich Ebert Stiftung. https://croatia.fes.de/filead min/user_upload/The_european_green_deal.pdf.

Murdiyarso, D., Dewi, S., Lawrence, D. and Seymour, F. (2011). *Indonesia's Forest Moratorium: A Stepping Stone to Better Forest Governance?* Working Paper 76. Bogor: Center for International Forestry Research (CIFOR). www .cifor.org/publications/pdf_files/WPapers/WP-76Murdiyarso.pdf.

National Climate Change Secretariat (2020). Green Growth Opportunities. www.nccs.gov.sg/singapores-climate-action/overview/green-growth-opportunities.

Ng, W. H. (2012). *Singapore, the Energy Economy: From the First Refinery to the End of Cheap Oil, 1960–2010*. London: Routledge.

Ng, R. J. (2020). ExxonMobil's New Tech Deployment in Singapore to Create 135 New Jobs, Says Teo Chee Hean. *The Business Times*, March 31. www .businesstimes.com.sg/energy-commodities/exxonmobil%E2%80% 99snew-tech-deployment-in-singapore-to-create-135-new-jobs-says-teo.

O'Gorman, R. (2020). Climate Law in Ireland: EU and National Dimensions. In D. Robbins, D. Torney and P. Brereton (eds.), *Ireland and the Climate Crisis*. London: Palgrave Macmillan, 73–89.

Oberthür, S. and Dupont, C. (2011). The Council, the European Council and International Climate Policy: From Symbolic Leadership to Leadership by Example. In R. Wurzel and J. Connelly (eds.), *The European Union as a Leader in International Climate Change Politics*. Abingdon: Routledge, 74–91.

Oberthür, S. and Roche Kelly, C. (2008). EU Leadership in International Climate Policy: Achievements and Challenges. *The International Spectator*, 43(3), 35–50. https://doi.org/10.1080/03932720802280594.

Olsson, D., Öjehag-Pettersson, A. and Granberg, M. (2021). Building a Sustainable Society: Construction, Public Procurement Policy and "Best Practice" in the European Union. *Sustainability*, 13(13), 7142. https://doi .org/10.3390/su13137142.

Oltermann, P. (2021). Joint CO2 Targets Must not Diminish German Industry, CDU Leader Warns EU. *The Guardian*, July 9. www.theguardian.com/world/ 2021/jul/09/joint-co2-targets-must-not-diminish-german-industry-cdu- leader-warns-eu.

Palmer, J. R. (2015). How Do Policy Entrepreneurs Influence Policy Change? Framing and Boundary Work in EU Transport Biofuels Policy. *Environmental Politics*, 24(2), 270–287. https://doi.org/10.1080/09644016.2015.976465.

Papadopoulos, Y. (2010). Accountability and Multi-level Governance: More Accountability, Less Democracy? *West European Politics*, 33(5), 1030–1049.

Pianta, S. and Sisco, M. R. (2020). A Hot Topic in Hot Times: How Media Coverage of Climate Change is Affected by Temperature Abnormalities. *Environmental Research Letters*, 15(11), 1–9. https://doi.org/10.1088/1748- 9326/abb732.

Pierre, J. and Peters, B. G. (2004). Multi-level Governance and Democracy: A Faustian Bargain? In I. Bache and M. Flinders (eds.), *Multi-level Governance*. Oxford: Oxford University Press, 75–89.

Pinandita, A. (2020). Indonesia to Receive $56 Million from Norway for Reducing Emissions. *The Jakarta Post*, May 22. www.thejakartapost.com/

news/2020/05/22/indonesia-to-receive-56-millionfrom-norway-for-redu cing-emissions.html.

Pollack, M. A. (2003). *The Engines of European Integration: Delegation, Agency, and Agenda Setting in the EU*. Oxford: Oxford University Press.

Prime Minister's Office Singapore (2019). SM Teo Chee Hean at the Inauguration of Total Regional Headquarters. *Chelsa_Cher*, December 9. www.pmo.gov.sg/Newsroom/SM-Teo-Chee-Hean-at-the-Inauguration-of-TotalRegional-Headquarters.

Raitzer, D. A., Bosello, F., Tavoni, M. et al. (2016). Southeast Asia and the Economics of Global Climate Stabilization. *Asian Development Bank (ADB)*. www.adb.org/publications/southeast-asia-economics-global-climate-stabilization.

Rankin, J. (2019). "Our House is On Fire": EU Parliament Declares Climate Emergency. *The Guardian*, November 28. www.theguardian.com/world/2019/nov/28/eu-parliament-declares-climateemergency.

Rankin, J. (2020). Focus on Coronavirus Shows Need for Climate Law, Says EU Official. *The Guardian*, March 4. www.theguardian.com/environment/2020/mar/04/focus-coronavirus-showsneed-climate-law-says-eu-official-frans-timmermans.

Renaldi, E. and Wires (2019). "Dark Day for Human Rights": Indonesian President Appoints Opposition Leader to Cabinet. *ABC News*, October 23. www.abc.net.au/news/2019-10-23/indonesia-president-appoints-opposition leader-to-cabinet/11631518.

Rietig, K. (2014). Reinforcement of Multilevel Governance Dynamics: Creating Momentum for Increasing Ambitions in International Climate Negotiations. *International Environmental Agreements: Politics, Law and Economics*, 14(4), 371–389. https://doi.org/10.1007/s10784-014-9239-4.

Rietig, K. (2019). The Importance of Compatible Beliefs for Effective Climate Policy Integration. *Environmental Politics*, 28(2), 228–247. https://doi.org/10.1080/09644016.2019.1549781.

Rietig, K. (2020). Multilevel Reinforcing Dynamics: Global Climate Governance and European Renewable Energy Policy. *Public Administration*, 99(1), 55–71. https://doi.org/10.1111/padm.12674.

Ritchie, H. (2019). Who has Contributed Most to Global CO2 Emissions? *Our World in Data*, October 1. https://ourworldindata.org/contributed-most-glo bal-co2.

Rodan, G. (2018). *Participation without Democracy: Containing Conflict in Southeast Asia*. Ithaca, NY: Cornell University Press.

Roger, C. (2020). *The Origins of Informality: Why the Legal Foundation of Global Governance are Shifting, and Why It Matters*. Oxford: Oxford University Press.

Ruysschaert, D. and Salles, D. (2016). The Strategies and Effectiveness of Conservation NGOs in the Global Voluntary Standards: The Case of the Roundtable on Sustainable Palm-Oil. *Conservation and Society*, 14(2), 73–85. https://doi.org/10.4103/0972-4923.186332.

Sadeleer, N. (2014). *EU Environmental Law and the Internal Market*. Oxford: Oxford University Press.

Sartor, O., Voss-Stemping, J., Berghmans, N., Vallejo, L. and Levaï, D. (2019). *Raising and Strengthening EU Climate Ambition: Priorities and Options for the Next Five Years*. Paris: Institute for Sustainable Development and International Relations (IDDRI). www.iddri.org/sites/default/files/PDF/Publications/Catalogue%20Iddri/Etude/201903-ST0119-raising%20EU%20ambition_0.pdf.

Scharpf, F. (1994). Community and Autonomy: Multi-level Policy-making in the European Union. *Journal of European Public Policy*, 1, 219–242. https://doi.org/10.1080/13501769408406956.

Schreurs, M. A. and Tiberghien, Y. (2007). Multi-Level Reinforcement: Explaining European Union Leadership in Climate Change Mitigation. *Global Environmental Politics*, 7(4), 19–46. https://doi.org/10.1162/glep.2007.7.4.19.

Seah, S. and Martinus, M. (2021). Gaps and Opportunities in ASEAN's Climate Governance. ISEAS – Yusof Ishak Institute. www.iseas.edu.sg/wp-content/uploads/2021/03/TRS5_21.pdf.

SG Climate Rally (2020). Climate Scorecard. https://scorecard.sgclimaterally.com/.

Shapiro, M. (2004). Deliberative, Independent Technocracy v. Democratic Politics. *Law & Contemporary Problems*, 68(1), 341–356.

Sinclair, L. (2020). The Power of Mining MNCs: Global Governance and Social Conflict. In: J. Mikler and K. Ronit (eds.), *MNCs in Global Politics: Pathways of Influence*. Cheltenham: Edward Elgar, 139–158.

Sirtori-Cortina, D., Afanasieva, D. and Dahrul, F. (2022). Nestlé Says It Will Drop Palm Oil Supplier Accused of Abuses. *Bloomberg*, September 29. www.bloomberg.com/news/articles/2022-09-29/nestle-to-drop-palm-oil-supplier-astra-agro-lestari-on-allegations-of-abuses.

Skjærseth, J. B. (2017). The European Commission's Shifting Climate Leadership. *Global Environmental Politics*, 17(2), 84–104.

Skjærseth, J. B. (2021). Towards a European Green Deal: The Evolution of EU Climate and Energy Policy Mixes. *International Environmental Agreements: Politics, Law and Economics*, 21, 25–41. https://doi.org/10.1007/s10784-021-09529-4.

Skjærseth, J. B. and Wettestad, J. (2010). Making the EU Emissions Trading System: The European Commission as an Entrepreneurial Epistemic Leader.

Global Environmental Change, 20(2),314–321. https://doi.org/10.1016/j.gloenvcha.2009.12.005.

Skovgaard, J. (2014). EU Climate Policy after the Crisis. *Environmental Politics*, 23(1), 1–17. https://doi.org/10.1080/09644016.2013.818304.

Sørensen, E. and Torfing, J. (2009). Making Governance Networks Effective and Democratic through Metagovernance. *Public Administration*, 87(2), 234–258. https://doi.org/10.1111/j.1467-9299.2009.01753.x.

SunStar (2019). Bacolod First City in PH to Declare Climate Emergency. July 20. www.sunstar.com.ph/article/1815146.

Tallberg, J. (2011). The Agenda-Shaping Powers of the EU Council Presidency. *Journal of European Public Policy*, 10(1), 1–19. https://doi.org/10.1080/1350176032000046903.

Tan, A. (2019). Parliament: About 75% of Industrial Emissions Are from Refining and Petrochemicals Sector. *The Straits Times*, October 7. www.straitstimes.com/politics/parliament-about-75-of-industrial-emissions-are-from-refining-and-petrochemicals-sector.

Tan, A. (2020). Emerging Stronger Taskforce Focused on Economic Recovery. *The Straits Times*, September 14. www.straitstimes.com/singapore/environment/emerging-stronger-taskforcefocused-on-economic-recovery.

Tan, S.-A. (2020). Singapore Non-Oil Domestic Exports Fall by 4.5% in May, First Decline after Three Months of Expansion. *The Straits Times*, June 17. www.straitstimes.com/business/economy/singapores-non-oil-domesticexports-fall-by-45-in-may-first-decline-after-three.

Taylor, K. (2021). EU Commission Clashes with Parliament over "net" 2030 Climate Target. *EURACTIV*, March 5. www.euractiv.com/section/energy-environment/news/eu-commission-clashes-with-parliament-over-net-2030-climate-target/.

Torney, D. (2015). *European Climate Leadership in Question: Policies towards China and India*. Cambridge, MA: MIT Press.

Tsafos, N. (2020). *Why Europe's Green Deal Still Matters*. Washington, DC: Center for Strategic and International Studies. www.csis.org/analysis/why-europes-green-deal-still-matters.

United Nations Development Programme (UNDP) and University of Oxford (2021). *Peoples' Climate Vote – Results*. UNDP and University of Oxford Department of Sociology. www.undp.org/publications/peoples-climate-vote.

United Nations Environment Programme (UNEP) (2021). State of the Climate. November 9. www.unep.org/explore-topics/climate-action/what-we-do/climate-action-note/state-of-climate.html?gclid=Cj0KCQjw-fmZBhDtARIsAH6H8qh dfB9HhLB24RjYfEi6egbR7vRDd2fBGt9iaZEMZPjHDiJOxtb4sGYaAijxEAL w_wcB.

UNFCCC (2019). Study on Cooperative MRV as a Foundation for a Potential Regional Carbon Market Within ASEAN. Synthesis Report. https://unfccc .int/page/publication-of-study-on-cooperative-mrv-as-a-foundation-for-apo tential-regional-carbon-market.

Van Schaik, L. and Schunz, S. (2012). Explaining EU Activism and Impact in Global Climate Politics: Is the Union a Norm- or Interest-Driven Actor? *Journal of Common Market Studies*, 50(1), 169–186. https://doi.org/10.1111/ j.1468-5965.2011.02214.x.

Varkkey, H. (2015). *The Haze Problem in Southeast Asia: Palm Oil and Patronage*. 1st ed. London: Routledge.

Varkkey, H. (2019). Winds of Change in Malaysia: The Government and the Climate. *HeinrichBöll-Stiftung Southeast Asia*, February 27. https:// th.boell.org/en/2019/02/27/winds-change-malaysia-government-and-climate.

Victor, D. G. (2009). *The Politics of Fossil-Fuel Subsidies*. Working Paper. Geneva: Global Subsidies Initiative of International Institute for Sustainable Development. www.ssrn.com/abstract=1520984.

Vietnam Plus (2022). Indonesia to Halt Renewable Energy Exports. June 7. https://en.vietnamplus.vn/indonesia-to-halt-renewable-energy-exports/ 229800.vnp.

Von der Leyen, U. (2019). Press Remarks by President von der Leyen on the Occasion of the Adoption of the European Green Deal Communication. *European Commission Press Corner*, December 11. https://ec.europa.eu/ commission/presscorner/detail/en/speech_19_6749.

Waldholz, R. (2019). 'Green wave' vs Right-Wing Populism: Europe Faces Climate Policy Polarisation. *Clean Energy Wire*, June 5. www.cleanenergy wire.org/news/green-wave-vs-right-wing-populism-europe-faces-climate-policy-polarisation.

Walker, H. and Biedenkopf, K. (2018). The Historical Evolution of EU Climate Leadership and Four Scenarios for Its Future. In S. Minas and V. Ntousas (eds.), *EU Climate Diplomacy: Politics, Law and Negotiations*. Abingdon: Routledge, 33–46.

Walton, K. (2019). Indonesia Should Put More Energy Into Renewable Power. *The Interpreter*, August 19. www.lowyinstitute.org/the-interpreter/indone sias-should-put-more-energy-renewable-power.

Westerwinter, O., Abbott, K. W. and Biersteker, T. (2021). Informal Governance in World Politics. *The Review of International Organizations*, 16(1), 1–27. https://doi.org/10.1007/s11558-020-09382-1.

Wicaksono, A. (2015). Energy Reform in Indonesia: One Year After the New Government. *Talk at the University of Queensland*, October 13.

Wijaya, A., Chrysolite, H., Ge, M. et al. (2017). *How Can Indonesia Achieve Its Climate Change Mitigation Goal? An Analysis of Potential Emissions Reductions from Energy and Land-use Policies*. Working Paper. Washington, DC: World Resources Institute.

Wijaya, T. and Nursamsu, S. (2020). The Trouble with Indonesia's Infrastructure Obsession. *The Diplomat*, January 9. https://thediplomat .com/2020/01/thetrouble-with-indonesias-infrastructure-obsession/.

Wijedasa, L. S., Sloan, S., Page, S. E. et al. (2018). Carbon Emissions from South-East Asian Peatlands Will Increase despite Emission Reduction Schemes. *Global Change Biology*, 24(10), 4598–4613.

World Bank (n.d.). World Bank Open Data. https://data.worldbank.org/.

World Meteorological Organization (WMO). (2022). WMO Global Annual to Decadal Climate Update. April 2022. https://hadleyserver.metoffice.gov.uk/ wmolc/WMO_GADCU_2022-2026.pdf.

Yeo, B. Y. (2018). COP-24, United Nations Climate Change Conference, December 12, 2018, Katowice, Poland. Speech at High Level Segment at COP-24.

Yukawa, T. (2018). European Integration through the Eyes of ASEAN: Rethinking Eurocentrism in Comparative Regionalism. *International Area Studies Review*, 21(4), 323–339. https://doi.org/10.1177/2233865918808035.

Zahariadis, N. (1992). To Sell or Not to Sell? Telecommunications Policy in Britain and France. *Journal of Public Policy*, 12(4), 355–376.

Zahariadis, N. (2003). *Ambiguity and Choice in Public Policy: Political Decision Making in Modern Democracies*. Washington, DC: Georgetown University Press.

Zahariadis, N. (2007). The Multiple Streams Framework: Structure, Limitations, Prospects. In P. A. Sabatier (ed.), *Theories of the Policy Process*, 2nd ed. Boulder, CO: Westview Press, 69–92.

Zahariadis, N. (2008). Ambiguity and Choice in European Public Policy. *Journal of European Public Policy*, 15(4), 514–530. https://doi.org/ 10.1080/13501760801996717.

Zohlnhöfer, R., Herweg, N. and Rüb, F. (2015). Theoretically Refining the Multiple Streams Framework: An Introduction. *European Journal of Political Research*, 54(3), 412–418. https://doi.org/10.1111/1475-6765.12102.

Zygierewicz, A. (2016). Implementation of the Energy Efficiency Directive (2012/27/EU): Energy Efficiency Obligation Schemes. *European Parliamentary Research Service*. www.europarl.europa.eu/RegData/etudes/ STUD/2016/579327/EPRS_STU(20 16)579327_EN.pdf.

Cambridge Elements ≡

Organizational Response to Climate Change

Aseem Prakash

University of Washington

Aseem Prakash is Professor of Political Science, the Walker Family Professor for the College of Arts and Sciences, and the Founding Director of the Center for Environmental Politics at University of Washington, Seattle. His recent awards include the American Political Science Association's 2020 Elinor Ostrom Career Achievement Award in recognition of "lifetime contribution to the study of science, technology, and environmental politics," the International Studies Association's 2019 Distinguished International Political Economy Scholar Award that recognizes "outstanding senior scholars whose influence and path-breaking intellectual work will continue to impact the field for years to come," and the European Consortium for Political Research Standing Group on Regulatory Governance's 2018 Regulatory Studies Development Award that recognizes a senior scholar who has made notable "contributions to the field of regulatory governance."

Jennifer Hadden

University of Maryland

Jennifer Hadden is Associate Professor in the Department of Government and Politics at the University of Maryland. She conducts research in international relations, environmental politics, network analysis, non-state actors, and social movements. Her research has been published in various journals, including the *British Journal of Political Science, International Studies Quarterly, Global Environmental Politics, Environmental Politics,* and *Mobilization.* Dr. Hadden's award-winning book, *Networks in Contention: The Divisive Politics of Global Climate Change,* was published by Cambridge University Press in 2015. Her research has been supported by a Fulbright Fellowship, as well as grants from the National Science Foundation, the National Socio-Environmental Synthesis Center, and others. She held an International Affairs Fellowship from the Council on Foreign Relations for the 2015–2016 academic year, supporting work on the Paris Climate Conference in the Office of the Special Envoy for Climate Change at the U.S. Department of State.

David Konisky

Indiana University

David Konisky is Professor at the Paul H. O'Neill School of Public and Environmental Affairs, Indiana University, Bloomington. His research focuses on US environmental and energy policy, with particular emphasis on regulation, federalism and state politics, public opinion, and environmental justice. His research has been published in various journals, including the *American Journal of Political Science, Climatic Change, the Journal of Politics, Nature Energy,* and *Public Opinion Quarterly.* He has authored or edited six books on environmental politics and policy, including *Fifty Years at the US Environmental Protection Agency: Progress, Retrenchment and Opportunities* (Rowman & Littlefield, 2020, with Jim Barnes and John D. Graham), *Failed Promises: Evaluating the Federal Government's Response to Environmental Justice* (MIT Press, 2015), and *Cheap and Clean: How Americans Think about Energy in the Age of Global Warming* (MIT Press, 2014, with Steve Ansolabehere). Konisky's research has been funded by the National Science Foundation, the Russell Sage Foundation, and the Alfred P. Sloan Foundation. Konisky is currently co-editor of *Environmental Politics.*

Matthew Potoski

UC Santa Barbara

Matthew Potoski is a Professor at UCSB's Bren School of Environmental Science and Management. He currently teaches courses on corporate environmental management, and his research focuses on management, voluntary environmental programs, and public policy. His research has appeared in business journals such as *Strategic Management Journal, Business Strategy and the Environment, and the Journal of Cleaner Production,* as well as public policy and management journals such as *Public Administration Review* and *the Journal of Policy Analysis and Management.* He co-authored *The Voluntary Environmentalists* (Cambridge, 2006) and *Complex Contracting* (Cambridge, 2014; the winner of the 2014 Best Book Award, American Society for Public Administration, Section on Public Administration Research) and was co-editor of *Voluntary Programs* (MIT, 2009). Professor Potoski is currently co-editor of the *Journal of Policy Analysis and Management* and the *International Public Management Journal.*

About the Series

How are governments, businesses, and non-profits responding to the climate challenge in terms of what they do, how they function, and how they govern themselves? This series seeks to understand why and how they make these choices and with what consequence for the organization and the eco-system within which it functions.

Cambridge Elements ≡

Organizational Response to Climate Change

Elements in the Series

Printed in the United States
by Baker & Taylor Publisher Services

Printed in the United States
by Baker & Taylor Publisher Services